W.Drawn

W9-CKE-386

Women Plantation Workers

Cross-Cultural Perspectives on Women

General Editors: Shirley Ardener and Jackie Waldren,
for The Centre for Cross-Cultural Research on Women, University of Oxford

ISSN: 1068-8536

Vol. 1:   *Persons and Powers of Women in Diverse Cultures*
Edited by Shirley Ardener

Vol. 2:   *Dress and Gender: Making and Meaning*
Edited by Ruth Barnes and Joanne B. Eicher

Vol. 3:   *The Anthropology of Breast-Feeding: Natural Law or Social Construct*
Edited by Vanessa Maher

Vol. 4:   *Defining Females: The Nature of Women in Society*
Edited by Shirley Ardener

Vol. 5:   *Women and Space: Ground Rules and Social Maps*
Edited by Shirley Ardener

Vol. 6:   *Servants and Gentlewomen to the Golden Land: The Emigration of Single Women to South Africa, 1820–1939*
Edited by Cecillie Swaisland

Vol. 7:   *Migrant Women: Crossing Boundaries and Changing Identities*
Edited by Gina Buijs

Vol. 8:   *Carved Flesh/Cast Selves: Gendered Symbols and Social Practices*
Edited by Vigdis Broch-Due, Ingrid Rudie and Tone Bleie

Vol. 9:   *Bilingual Women: Anthropological Approaches to Second Language Use*
Edited by Pauline Burton, Ketaki Dyson and Shirley Ardener

Vol. 10: *Gender, Drink and Drugs*
Edited by Maryon MacDonald

Vol. 11: *Women and Mission: Past and Present*
Edited by Fiona Bowie, Deborah Kirkwood and Shirley Ardener

Vol. 12: *Women in Muslim Communities: Religious Belief and Social Realities*
Edited by Camillia Fawzi El-Solh and Judy Mabro

Vol. 13: *Women and Property, Women as Property*
Edited by Renée Hirschon

Vol. 14: *Money-Go-Rounds: Women's Use of Rotating Savings and Credit Associations*
Edited by Shirley Ardener and Sandra Burman

Vol. 15: *'Male' and 'Female' in Developing Southeast Asia*
Edited by Wazir Jahan Karim

Vol. 16: *Women Wielding the Hoe: Lessons from Rural Africa for Feminist Theory and Development Practice*
Edited by Deborah Fahy Bryceson

Vol. 17: *Organizing Women: Formal and Informal Women's Groups in the Middle East*
Edited by Dawn Chatty and Annika Rabo

MONTGOMERY COLLEGE
GERMANTOWN CAMPUS LIBRARY
GERMANTOWN, MARYLAND

Women Plantation Workers

International Experiences

Edited by
Shobhita Jain and Rhoda Reddock

Oxford • New York

251722
DEC 10 2000

First published in 1998 by
Berg
Editorial offices:
150 Cowley Road, Oxford, OX4 1JJ, UK
70 Washington Square South, New York, NY 10012, USA

© Shobhita Jain and Rhoda Reddock 1998

All rights reserved.
No part of this publication may be reproduced in any form
or by any means without the written permission of Berg

Berg is the imprint of Oxford International Publishers Ltd.

Library of Congress Cataloging-in-Publication Data

A catalogue record for this book is available from the Library of Congress.

British Library Cataloguing-in-Publication Data

A catalogue record for this book is available from the British Library.

ISBN 1 85973 972 5 (Cloth)
     1 85973 977 6 (Paper)

Typeset by JS Typesetting, Wellingborough, Northants.
Printed in the United Kingdom by Biddles Ltd, Guildford and King's Lynn.

Contents

Acknowledgements vii

Preface
Sidney Mintz ix

Notes on Contributors xv

1 Plantation Women: An Introduction
  Rhoda Reddock and Shobhita Jain 1

2 Women Field Workers in Jamaica During Slavery
  Lucille Mathurin-Mair 17

3 The Indentureship Experience: Indian Women in Trinidad
  and Tobago 1845–1917
  Rhoda Reddock 29

4 Migration, Labour and Plantation Women in Fiji:
  A Historical Perspective
  Shaista Shameem 49

5 Tamil Women on Sri Lankan Plantations: Labour Control
  and Patriarchy
  Rachel Kurian 67

6 Indian Migrant Women and Plantation Labour in Nineteenth- and
  Twentieth-century Jamaica: Gender Perspectives
  Verene A. Shepherd 89

7 Gender Relations and the Plantation System in Assam, India
  Shobhita Jain 107

8 Women's Role in the Household Survival of the Rural Poor:
  The Case of the Sugar-cane Workers in Negros Occidental
  Violeta Lopez-Gonzaga 129

9 Women Plantation Workers and Economic Crisis in Cameroon
  Piet Konings 151

Bibliography 167

Index 177

Acknowledgements

We would like to take this opportunity to thank Gloria Lawrence for typing, and Olive Seaton for copy editing, the first draft of the manuscript. Alyson Myses and Lyndon Guisseppi provided invaluable research assistance to Rhoda and we are also thankful for the small, but useful, assistance provided by the Campus Research Fund of the University of the West Indies, St. Augustine.

Geeta Sharma kindly took on the typing of the final version of the manuscript and Judy Mabro copy edited this with her usual high standard and eye for detail.

We would also like to thank both Shirley Ardener and Jackie Waldren for evaluating the manuscript for publication, and Sidney Mintz for writing the preface.

The Centre for Cross-Cultural Research on Women, Queen Elizabeth House, Oxford University, awarded Shobhita Jain a Visiting Research Fellowship and Overseas Development Administration, London, financed it. We acknowledge with thanks the support this provided for preparing a hard copy of the manuscript for Berg Publishers at Oxford.

Shobhita Jain
Rhoda Reddock

Preface

The writers of this book have all based their work on the axis of interaction of two subjects. On the one hand, they write about women. On the other, they write about plantations. Plantations of the sort they describe can be dated more or less exactly. They typify an epoch that had its beginnings around 1550, but in some ways has not yet quite ended. During the last four centuries or so, the evolution of agrarian institutions such as the plantation first took place as national economies – the economies of Spain, England and France, for example – were being consolidated and becoming international, by acquiring overseas colonies. But that evolution continued, as a global economy was becoming more visible and important, eventually even beginning to supplant or displace national economies. Institutions that were parts of the national economies, including those in the colonies, had definable economic and social forms, of which the plantation was one. By concentrating on the experiences of women plantation workers, then, these writers are also identifying a particular, epochal, sort of experience in the history of one-half of humanity.

The contributions to this book aim to provide the reader with a sweeping overview of the fates, past and present, of female plantation labourers. The plantation itself is a special form of large estate. Most are located in subtropical zones, are worked by masses of usually unskilled labourers, and are engaged in the production of basic commodities (foods, industrial staples) for overseas markets. They have been around, in more or less the same form, for more than four centuries.

Plantations have bad reputations; they are linked to colonialism, to the economic and sexual exploitation of non-white peoples, and to exhausting and badly-paid manual labour. Their history in relation to the status of women is a special chapter, but until now relatively little is available on the subject for the general reader. In the chapters of this book the specially disadvantaged position of female workers, and the reasons for their degraded status, are laid bare. But the writers also paint a stirring picture of how women tried to resist their condition, how they struggled to cope with the circumstances in which they found themselves.

Three chapters have to do with the New World, and describe the situation of enslaved or indentured (contracted) female sugar-cane labourers – they deal primarily with the past. Six concern the Old World, with women working on plantations that produce sugar-cane in two instances (Fiji, the Philippines), tea in three others (India, Sri Lanka, Cameroon), rubber and several crops (including oil palm and tobacco) in the last (Sumatra, Indonesia). Some deal with the recent past, but others focus on the present.

In their introduction, the editors of this volume examine all of the contributions, one by one. Hence this prefacer's task is made simple: to indicate in a general manner what purpose a collection of this sort serves, and where it might stand in relation to existing literature on the subject. Even this task, however, requires an historical perspective, for the plantations of today can be adequately under-stood in the light of the four centuries of modern existence. Large agricultural estates, after all, are truly ancient; there are descriptions of the latifundia in the Roman empire. But during the sixteenth-century transition to capitalism, plantations of the post-Columbian sort came to typify a lengthy and special era of 'develop-ment': they were operated by Europeans, in colonial areas, in order to produce basic staples for the West, using unskilled non-white, generally imported, coerced or contracted labour. Given the fact that their labour was mostly not free, they could by no means be called typical capitalist enterprises; but they became integral parts of the world economy, and their profits were invariably drained back to the metropolises for capitalistic purposes. How such enterprises took shape is a fundamental part of this story.

Post-Columbian Era

To remark on how much the world has changed since the so-called 'Columbian' or fifteenth century is to bandy platitudes. From the point of view of the present collection, however, the Columbian achievement was a special turning-point, because of its long-range effects on world agriculture and mass consumption. It resulted in the transfer to the American subtropics of agrarian forms, labour arrangements and crops long familiar in the Old World, which would subsequently flourish to a degree undreamed of before 1492. At a later time, transformed, 'modern' plantations would re-expand in the Old World as well.

From what is now called the United States South to the southern reaches of today's Brazil and Peru, and centring upon the Antillean islands; the most important (though not only) labour arrangement was slavery; and the most important crop (though again, not exclusively) was sugar-cane.

The plantation system, as described earlier, was already operating in the eastern Mediterranean and elsewhere, long before the discovery of the Americas. But until after the discovery of the New World, there did not exist the aggregate market forces, and the state and entrepreneurial pressures, which would vastly increase the importance of plantations over time. For the most part plantations lay in the subtropical and tropical zones; they were established within the colonies of the

European powers; and the goals of their production were the growing markets of the metropolises. The sugar, rum, molasses, tobacco, indigo and cotton (and soon enough other staples besides) which they produced came to be used, more and more to make commodities for 'ordinary' people of the Western nations, whose rising buying power under expanding capitalism enabled them to afford them. Though the large-estate form could be used in the production of many different substances and objects, it had long been most important for the production of sugar.[1]

Slavery as a labour form is, of course, ancient. In the New World, however, it became 'industrial' in nature, which is to say highly disciplined and commercialized (rather primarily domestic), depersonalized (organized by squads of workers, rather than by individual skills), and extremely large in scale. During nearly four centuries as an institution, New World slavery led to the importation of almost ten million Africans to the Americas who, with their more numerous descendants, powered the plantation system throughout a vast intermediate zone linking the Americas from north to south.

By the middle of the nineteenth century, due to a variety of powerful pressures (not the least being the resistance of the slaves themselves), several of the leading slave powers, such as Britain and France, had given up slavery, and the search for alternate labour, working under different arrangements, had begun. It was owing to that search that contract labour arrangements replacing slavery were installed, and the movement of contracted Chinese Indians, Africans, Sãotomeños, Canary Islanders, Javanese and others to the plantation zones of the New World had begun. The nineteenth century also saw the establishment of new plantations in the Old World – Mauritius, Fiji, South Africa and elsewhere – and their introduction, in new form, into Old World sugar-producing regions, such as Java and Taiwan; and in India and Sri Lanka (formerly Ceylon), for the production of tea. Nearly all of these new developments in the Old World also brought in their wake the migration of labourers from elsewhere.[2] Thereafter, the demand for agro-industrial materials such as palm oil, jute, copra and rubber led to a somewhat different era in plantation development, principally in Asia, but also in Africa. To this day the plantation system remains immensely important in world agriculture, even though slavery is now gone, and many plantation products have since been supplanted by synthetics.

Other Forms of Plantation Labour

Throughout the history of the plantation in both Old World and New, the theme of labour has always figured importantly and, often enough, tragically. The successive emancipations of the nineteenth century, intended to free humankind from hereditary bondage, were followed by the managed and actively encouraged inflow of other labourers, often from different parts of the same colonial world. W. Arthur Lewis

(1969) pointed out that, of the estimated hundred million persons who crossed oceans during the nineteenth century in search of work, half were going from one tropical or subtropical zone to another. Most of these people were moving from a colonial area (such as India) to another (such as Mauritius, British Guiana, or Fiji). Yet others, such as the Javanese, participated in several migratory waves. One such went from Java to Sumatra; another would one day be shipped from Java to Dutch Guiana (Surinam). It was typical that such labour would be sought either in colonies of the West, which lacked a political voice in the metropolis, such as India or Java; or in countries (such as China) too weak politically to defend their own citizens overseas. Millions of Indians, Chinese and other migrants were transferred in these movements, supposedly armed with 'contracts' to defend their rights; but over and over again it turned out that the planters were protected by the contracts, not the workers. The nineteenth century was the demographic outcome of the Industrial Revolution; but it was also the demographic harbinger of industrial modernity. That has, in turn, brought about yet another world-scale migration, to which we are all witness. But from the nineteenth-century perspectives of those in power, the transoceanic movement of a hundred million persons, with all of their attendant miseries, was merely a passing moment in the international redivision of labour. Migration, by choice or perforce, was how that redivision was achieved.

Of the contributions to this volume it is noteworthy that in only one (by Lopez-Gonzaga on the Philippines) are the workers *not* the descendants of migrants. But that the plantation system should have depended so consistently on migrant labour is actually not surprising. As Edgar Thompson (1959)[3] pointed out, the most important characteristic of the plantation is not the crop that it grows, its specific labour arrangements, or the climate that typifies it, but the fact that it is a *frontier institution*. Such enterprises usually engrossed large land areas (often after their inhabitants had been killed off, or died of disease), which they could then 'put to higher uses' by producing world market staples. From the first sugar plantations of the early sixteenth century until the third decade of the nineteenth century, plantation slavery enabled the planters to profit from an immobilized labour force that had been dragged to its labours in chains. In the nineteenth century, 'contract or indenture served the same objectives. In the twentieth, in most old plantation areas, local population has grown enough to make any form of coercion unnecessary – except, of course, the coercion of hunger.[4] The plantation form, then, exhibits both broad continuities with the past, and significant changes in the way it operates.

The history of colonial, erstwhile colonial or quasi-colonial regions provides a backdrop against which it is easier to see the more traditional disadvantages under which migrant labour struggled. The first such disadvantage may seem too obvious to mention: being colonies, the populations of those regions had little or no say in their economic fate. Thus, for example, in the British West Indies, such as Jamaica, Trinidad and Tobago and British Guiana (the now-independent countries of Jamaica,

Trinidad and Tobago and Guyana) in the nineteenth century, no freed person was ever consulted about the importation of migrant labour from elsewhere – even though immigrants were clearly being imported in order to drive down the price of the freed people's labour. In societies such as Jamaica, local legislatures levied taxes on the most necessary imports of the freed persons, such as lumber, cornmeal and dried fish, then used those taxes to import other labourers. In Guyana, robust efforts were made by the legislature, composed mostly of planters, to keep free people from access to agricultural land. In short, the colonial status of these places kept their populations powerless, and nullified any idea of democratic decision-making by use of the ballot.

The second such disadvantage flowed from the first: the inability of local people (migrant or not) to advance themselves economically, since so little was invested by the metropolis to change the economic conditions or opportunities of the colonies. This disprivileging of local people was continued, of course, in the fates of their children. The lack of opportunity which greeted migrant newcomers was inherited, as it were, so that successive generations faced the same conditions as their parents and grandparents. Inferior schools, inferior medical facilities, bad labour conditions and the absence of alternative employment, all kept the labour force immobile. It was not slavery, no; but it was like slavery in many of its effects.

Still another point deserves to be made in this same connection. A punitive colonial ambience was imposed on all of the people who worked on plantations, white and non-white, migrant and local, Christian and non-Christian. But it was imposed far more onerously upon women than upon men. Why this should have been the case is one of the major questions that the contributors to this volume address.

The relationship of gender to the division of labour is an issue as old as our species. But no humanistic discipline in Western thought, anthropology included, has ever been able to free its ideas entirely from the prevailing social system within which it functioned. Despite that reality, during the last several decades the study of gender has been lifted to a new plane in the social sciences and in history. At the same time, even the recent feminism of the West has suffered from a common inability to integrate its insights in a way that transcends effectively the differences in class and colour which so often divide people, and inflect their consciousness. It is not a new complaint; feminist leaders and intellectuals in the West have been drawn in their majority from among white, college-educated, professional (and thus highly privileged) women, for whom issues of status equality bulk much more tellingly than do issues of acceptance in other spheres, such as economic equality or political equality. To some degree, then the chapters that follow may both chill – and brace – dozing intellectuals. The sufferings of female plantation labourers may seem remote to most of us. But they are still very much part of the contemporary world. Their stories will arouse anger at arbitrary injustice. But they should also excite reflection. To the extent that people still suffer in these ways – and many do – we need to ask ourselves what we can do about it.

The contributions to this volume are not conclusive, nor were they intended to be. Readers will be afforded a close-up understanding of the female protagonists, who are given an opportunity to be heard. But the larger issue which the book reveals is that of the fates of people – particularly women and children, worldwide – whose economic privation not only exploits them, but also keeps their descendants exploited. One can hope that this book will be followed by yet others, concerned with the same theme. If that hope is realized, then the book will have served its purpose.

Sidney W. Mintz

Notes

1. A slave revolt involving thousands of East African agricultural labourers took place in the Tigris–Euphrates Delta in the mid-ninth century; and they may have been sugar-cane plantation workers (see Mintz 1985: 27; Popovic 1965). The sugar-cane plantation worked with slave labour remains to some extent an archetype. Though neither sugar nor slave labour have been emblematic of the plantation for more than a century, there is no doubt that the majority of slaves brought to the New World were brought there for that purpose.

2. Sugar – or rather the great commodity market which arose demanding it – has been one of the massive demographic forces in world history. Because of it, literally millions of enslaved Africans reached the New World, particularly the American South, the Caribbean and its littorals, the Guianas and Brazil. This migration was followed by those of East Indians, both Muslim and Hindu, Javanese, Chinese, Portuguese and many other peoples in the nineteenth century. It was sugar that sent East Indians to Natal, sugar that carried them to Mauritius and Fiji. Sugar brought a dozen ethnic groups in staggering succession to Hawaii, and sugar still moves people about the Caribbean. Though sugar is called the culprit here, other plantation crops resulted in like migrations, and of course 'sugar' did nothing – it was the nations and investors who brought the plantations into being who created the migrations.

3. In fact long ago in 1932 he wrote his Ph. D. dissertation in Sociology at the University of Chicago and an extract from it was published later in 1959; see Bibliography.

4. Hence today, it is not considered unusual for Jamaican labourers to be flown to Florida to cut sugar-cane during the harvest, only to be flown back to Jamaica thereafter. Instead of moving the plantations, modern capitalism simply moves the labour. But this case is somewhat unusual since it is situated geographically in a developed country, not a colony or recent colony (see Wilkinson 1989).

Notes on Contributors

Shobhita Jain Professor of Sociology at Indira Gandhi National Open University, New Delhi. Among her publications are *Sexual Equality: Workers in an Asian Plantation System* (1988); *Case Studies of Farm Forestry and Wasteland Development in Gujarat, India* (1988) and *Bharat Mein Parivar, Vivah aur Natedari* (1996).

Piet Konings Head of Department for Political and Historical Studies at the African Studies Centre at Leiden in the Netherlands. He has carried out extensive research on labour and trade unionism in Ghana and Cameroon. His books include *The State and Rural Class Formation in Ghana; Labour Resistance in Cameroon* (1986); *Itinéraires d'accumulation au Cameroun* (co-edited with Peter Geschiere, 1993) and *Gender and Class in the Tea Estates of Cameroon* (1995).

Rachel Kurian Institute of Social Studies, The Hague. The International Labour Organization published in 1982 her report, *Women Workers in the Sri Lanka Plantation Sector: An Historical and Contemporary Analysis* and the University of Amsterdam awarded her in 1989 a doctorate on her thesis, 'State, Capital and Labour in the Plantation Industry in Sri Lanka 1834–1984'.

Violeta Lopez-Gonzaga Social Research Centre, University of St. La Salle, Bacolod City, Philippines.

Lucille Mathurin-Mair Ministry of Foreign Affairs and Trade, Kingston, Jamaica.

Rhoda Reddock Head, Centre for Gender and Development Studies, and Senior Lecturer, Department of Sociology, The University of the West Indies, St. Augustine, Trinidad & Tobago. Among her publications are *Women, Labour and Politics in Trinidad and Tobago: A History* (1994) and *The Contemporary Caribbean* (co-edited with R. Deosaran and N. Mustapha in 1993).

Shaista Shameem Lecturer in Sociology and Social Anthropology at Waikato University, New Zealand. She holds a Ph. D. from Waikato University, where she is also working towards a law degree. Her academic interest are class, gender, ethnicity, Fijian political economy and Pacific legal systems and constitutions.

Verene A. Shepherd Senior Lecturer in the Department of History, the University of the West Indies, Mona, Jamaica. She is Secretary-Treasurer of the Association of Caribbean Historians and author of *Transients to Settlers: the Experience of Indians in Jamaica* and co-editor with Hilary Beckles of *Caribbean Slave Society and Economy and Caribbean Freedom* and with Barbara Bailey and Bridget Brereton of *Engendering History: Caribbean Women in Historical Perspectives*.

1

Plantation Women: An Introduction

Rhoda Reddock and Shobhita Jain

Why This Collection?

In discussions with scholars carrying out work on women in plantation societies, certain common characteristics emerged. While some of these were experienced by most groups of oppressed and superexploited women, there were also different ones which began to suggest a certain pattern.

The plantation being a 'total institution' (Goffman's concept applied by Best 1968) operated under conditions which gave rise to specific characteristics of the sexual division of labour. It also created new spaces for negotiated alternatives in male–female relationships. In addition, the plantation experience of initially coerced labour based on strict profit and loss calculations, provided indications of the productive capabilities of women which contemporary ideologies in other parts of the world were seeking to deny. Further, the plantation experience presents a clear example of the ways in which the sexual division of labour and gender roles can be differentially applied depending on the system of labour organization, the class, race, ethnicity or nationality of the workers involved.

The editors felt that there was a strong need to gather into one volume the few studies that have been done on female plantation labour. A collection like this could help to make the data on plantation women more easily available and accessible. It is hoped that this will encourage further research and exchange of ideas, as well as the emergence of a broader framework for analysis. In the existing literature on plantation labour, the role of female labour has found very little space. This omission is particularly conspicuous given the large number of women labourers employed in many plantation regimes. Covering widely situated plantation territories during slavery to colonial to post-colonial eras, this collection aims to fill a major gap in the existing literature. For a collection of essays on capitalist plantations in colonial Asia, see the special issue of the *Journal of Peasant Studies*, titled 'Plantations, Proletarians and Peasants in Colonial Asia', edited by Daniel et al. 1992. Another publication, *Women in Colonial India: Essays on Survival, Work and State*, edited by Krishnamurty (1989) notes 'survival strategies' in the plantation sector (e.g. Lal 1989). However, neither publication allows sufficient space to female labour on plantations, though one is about plantations and the other is about women. The context of colonial or neocolonial and capitalist discourse cannot afford to overlook constructions of gendered positions.

The adaptability of the plantation economies from slavery to colonial to post-colonial regimes shows that the plantation mode of production is to continue in all those countries where state enterprises have failed to bolster economic growth (see Graham and Floering 1984). Moreover, social, economic and cultural contacts between the plantation enclaves and their surroundings have brought about certain changes in societies that are penetrated by the plantation system. To understand the changes, socio-economic generalities necessary for perpetuation of the plantation system, need to be criss-crossed with the cultural patterns of the plantation population. Aspects of local culture, of the culture of migrant labourers, are equally powerful constraints. This book aims to make the shift from socio-economic gener-alities to understanding cultural specificities. Let us first consider the internal structures of the plantation.

The Origins of the Plantation

In its early usage, the term 'plantation' was often interchangeable with 'colony'. European colonization created these new societies as a part of their economic and physical expansion into previously unexploited territories. The plantation system of agriculture has in fact been the classical form of capitalistic exploitation in tropical areas where it developed as a political and social as well as an economic institution.

The term plantation was originally used to designate a plot of ground set with plants. During the period of British colonization of the West Indies and North America it came to denote a group of settlers as well as the political unit constructed by such a group (Selyman and Johnson 1948). Although plantations are usually located in tropical areas, this is the only way in which they are tropical. The ownership and investment traditionally comes from countries in the North. Local labour is utilized but local ownership was until recently, extremely rare. The plantations fulfil a desire on the part of non-tropical peoples for objects which can only be provided by tropical lands.

Plantations in the West Indies developed sugar as their principal crop in the sixteenth and seventeenth centuries. It was as sugar estates that the West Indies and mainland territories first became of value to the British, the French, the Dutch and to a lesser extent the Spanish. Sugar was known in Europe as the 'honey-bearing reed of India' (Greaves 1959: 13). Although coffee, cotton and indigo were among the first crops planted by British and French settlers in the West Indies, they were soon overshadowed by the sugar-cane. Sugar was brought to Suriname and the West Indian islands by the Jewish settlers who had gone to Pernambuco during its occupation by the Dutch. But, as pointed out by Mintz (in a personal communication):

> The idea that the first sugarcane grown and the first sugar made in the New World was in Brazil is quite wrong. Sugarcane did not diffuse from Pernambuco to the Antilles. It is true that the Dutch taught the Barbadian colonists to produce sugar, but that was more than a century after it was first grown in Santo Domingo. Sugarcane was brought to the New World on the second voyage of Columbus. It was cultivated on the Spanish island of Santo Domingo, and the first cask of sugar in the New World was made from it and shipped to Spain in (or about) 1516. Towards the end of the sixteenth century the Spanish also took the cane from Pernambuco and planted it in Puerto Rico.

In China and Japan tea was an indigenous crop long before it became a plantation crop for export. However, from the middle of the nineteenth century, cultivation of the crop in India, Ceylon and Java was solely for export purposes and the plantation system was responsible for expansion in the trade. Sugar-cane was also widely grown and crudely processed in these areas long before it was assimilated as a plantation crop.

Cocoa was introduced into Europe in the early part of the seventeenth century from Mexico. Today, West Africa rivals Central America as a major source of cocoa. Cocoa was first transferred to West Africa by the establishment of cocoa plantations on the island of San Thome by the Portuguese and also on the island of Fernando Po by the Spanish. Seeds were taken from Fernando Po in the nineteenth century by West African labourers returning to their homeland.

Bananas and rubber can perhaps be viewed as the plantation products of the twentieth century as sugar was of the seventeenth and cotton of the nineteenth. The successful export of bananas depended on the improved sea transportation of the twentieth century, so it was only then that the trade expanded.

In a unique manner the plantation system at a global level has been responsible for the merging of enterprise, capital and labour from different parts of the world to work together in areas which offer new opportunities for tropical crop production.

In the West Indies labour for the plantations originally came from the indigenous island peoples, the Caribs and the Arawaks/Tainos. However, initial efforts to mobilize the native population as a labour force failed because the Amerindians died in their thousands, through suicide or from being too weak to withstand the hard labour. The most successful solution found to the labour problem was the large-scale importation of African slaves. After the abolition of the slave trade in the nineteenth century and the emancipation of slaves shortly after, a new source of labour was sought. East Indians were brought to the plantations through a system of indentureship which was a type of enforced contract labour for fixed periods of time.

In India, China, the Straits Settlements and the Dutch East Indies, the large dependent labour force mainly consisted of Chinese, Javanese and Indians. These labourers were imported at the expense of the planters. Today labour is still of a forced nature despite the abolition of indentured labour. The pay for workers is low and the living conditions are poor.

The Plantation as a System: Change and Transition

The plantation is by definition a class structured system of organization, strongly hierarchical and male dominated in nature. The original plantations were rooted in a considerable degree of centralized control. In the 'old style' plantations, labour was bound, using some mechanism of outright coercion such as slavery, peonage or indentured servitude. Part of the resources of the enterprise – that is the labour time of the workers – was employed to underwrite the consumption needs of the workers and the status needs of the owner (Wolf and Mintz 1957). Labour did not feed itself outside of the boundaries of the plantation.

The 'new style' plantation, on the other hand, according to Wolf and Mintz (1957) is based on 'rational' cost accounting where the consumption needs of both owners and workers were no longer relevant to its operations. By the same token, labour costs are determined by competition or by other factors which affect this competition, such as labour organization. Some of the major characteristics of old style plantations are firstly a predominance of personal face-to-face relationships. Personal action is used to carry out technical functions, most importantly ritual relationships of dependence and dominance between workers and owners.

The old style plantations differ in a large measure from new style plantations in terms of their operational style and some of their objectives. New style plantations are, supposedly, characterized by impersonal relationships between owners and workers. According to this view, the plantation takes no responsibility for the employed labour force, in other words there is no paternalism involved in their relationship. New style plantations are said to be focused on rational efficiency as their production goal and this type of plantation is not an apparatus for servicing the status needs of its owners or managers. Nonetheless, the production process in the new style plantations was not immune to the cyclical fluctuations, experienced in the growth of capitalism during the last two hundred years. Production relations in most new style plantation regimes continued to feature coercion of the labour.

Courtenay (1980), on the other hand, developed a three-part typology of the traditional, the industrial and the modern plantation. Following Best (1968), the traditional plantation was defined as a total economic institution in which the entire existence of the workforce was incorporated into the process of production and reproduction. The inflexibility inherent in the system was tied up with the range of crops produced – sugar, cotton, tobacco, rice and indigo – and the need to extract the maximum labour from their investment. This form of production, of course, with significant variations, characterized the slave plantations of Brazil, the West Indies, the mid-Atlantic and Southern USA especially Virginia, the Carolinas and Georgia for a period of close to three hundred years. The traditional slave plantation provides the context for Lucille Mathurin-Mair's article on women field workers in Jamaica during slavery.

Changes in the plantation system in the nineteenth century were to a large extent wrought by developments in the international economy and in local production relations. One of the most important of these was the abolition of the British slave trade in 1807 and the eventual emancipation of slaves in British colonies between 1833 and 1838. These developments were the result both of the continued slave revolts and resistance in the colonies and the emergence of industrial capitalism in the metropole. The other important development was the expansion of the plantation system to new areas in South and South East Asia, Africa and the Pacific and the production of new crops (Courtenay 1980: 44–5).

The new industrial plantations which emerged were adapted to the new industrial era. No longer dependent on slave labour, bonded or indentured labour had to be introduced. Due to increasing competition and technological improvements there was a need to maximize economies of scale. This was effected by amalgamating smaller estates, mechanization, use of improved crop varieties and utilization of land use by, for example, reducing or eliminating the growth of food crops for estate consumption. All of this took place under the ownership of joint stock companies as opposed to the single ownership of the past. The still 'total' character of both the traditional and industrial plantations as we will see in the chapters of this book, presented the possibility of a challenge to traditional patterns of male authority and control. Women's work in agriculture was often as important as or more important than their work in biological or social reproduction (see Edholm, Harris and Young 1977); so much so that plantation labour continued to be primarily derived from migration and until this century the plantation management still had the responsibility to provide basic rations, housing, child care and health facilities for their labourers, albeit at the lowest possible level. Mechanized technology as a substitute for labour-intensive methods was not adopted in most of the new plantations where employers could get cheap and adequate labour supply.

The labour needs of the plantation system occasioned some of the most significant population movements in modern history. During this period indentured labour from the colonized parts of Africa, Asia and China were transported to various countries. By far the largest number were Indians or 'coolies' who were taken to plantations in the West Indies including British Guiana, Jamaica, Suriname,

Trinidad, Southern and East Africa, Fiji, Malaysia, Ceylon and within India itself to Assam. A number of studies based on this form of labour are included in this collection, including the chapters by Kurian on Ceylon (now Sri Lanka), Jain on the Assam tea gardens, Reddock on Trinidad and Tobago, Shameem on Fiji, and Shepherd on Jamaica focusing on the indentureship experience in the post-slavery plantation era.

Another characteristic of the industrial plantation was its coexistence with small farmer production. Small peasant production of estate crops was one way of reducing risk, ensuring the continued use of skilled labour especially female workers withdrawn from public labour on the estates. All processing, however, took place in the plantation factory.

As individual investment gave way to company ownership, investments by trading companies had to be safeguarded and hence often accompanied or followed political colonization by the European powers. By the nineteenth century a trading system had been established which facilitated the flow of raw materials and highly valued plantation crops to European manufactured goods, a situation which in many respects still exists today (Best 1968).

The modern plantation according to Courtenay is much less easily recognizable than its predecessors, largely because many of the characteristics previously unique to the plantation have now been adapted to other forms of agricultural enterprise. This fact and the continued existence of plantations even in their modern form are testimony to the economic advantages of the plantation as an instrument of accumulation. The modern plantations have rationalized land use and used scientific research to improve agricultural methods. This has resulted in raising labour productivity. All the same, labour control in these plantation regimes continues to centralize unfree labour, a feature of colonial plantations. This is the reason why women workers on tea estates in Cameroon are proving to be a problem to the management, as is shown by Piet Konings (1995a, b).

The historic plantations were significant players in the integration of global economic systems. Today, many transnational corporations, continuing to produce, process and market plantation crops, often find it difficult to take policy decisions in the wake of constantly changing politico-economic paradigms in developing countries (see Tiffen and Mortimore 1990). Those with more practical and less theoretical concerns like Graham (1984) argue that 'the modern plantation offers the option of using this as the most efficient way of utilizing available factors of production to provide a maximum social return'. But a review of the social impact of plantations and the issue of workers' conditions would belie the optimism generated by the rationalized character and function of the plantations in the contemporary world. All eight case-studies in this volume suggest that to consider plantations as enclaves in dual economies and their dependency on foreign politico-economic systems (as developed by Beckford 1972) remains a useful way to analyse the patterns of labour control and actual (as opposed to those mentioned in various labour conventions and legislations) living conditions of the plantation workers.

Sajhau and Von Muralt (1987) in their review of changes and developments in the plantation system over the period from 1965 to 1985 have also shown that 'most countries with plantations remain dependent on plantation crops as a major source of foreign exchange or raw materials, and the problems of plantation workers are essentially the same'. About women workers Sajhau and Von Muralt (1987: 120–21) say that:

> Attempts to differentiate between male and female labour meet with even greater obstacles which frequently result from the fact that family and not the individual is contracted for work, particularly when wages are based on a piece-rate system. The male–female ratio does not follow the same trend in different countries. Where there are labour shortages and better-educated men can find jobs in manufacturing, the percentage of women workers on plantations increases; on the contrary, where unemployment and underemployment are widespread, permanent jobs on plantations may be given to men, with women kept as a labour reserve for peak seasons. The types of crops grown also determine to a certain extent the sexual division of labour: sugar-cane and oil palm plantations usually employ fewer women than do tea and coffee plantations.

Our case-studies show that between one-third and a little more than a half of the plantation labour force are women. In 1981, 382,130 women labourers as against 379,192 men were employed on Indian tea plantations. This was also the case in Sri Lanka; slightly more than half of the 776,000 workers were women. Both in India and Sri Lanka, tea plantations employ the pluckers (essentially women) on a permanent basis. They are also required to work overtime in peak periods while men in such situations remain underemployed. In African countries the trend of women being fully occupied in farming family plots seems to be changing as Piet Konings has shown in the case of Cameroon. 'In most African countries the proportion of women working on plantations is nearing 50 per cent' (Sajhau and Von Muralt 1987: 122).

Plantations and Haciendas

In this book, plantations as discussed above have been the main focus. However, at least one contribution, by Lopez Gonzaga on the Philippines, describes a hacienda-type situation. It is important, therefore, to define clearly the characteristics – similarities and differences – of both these agricultural entities.

According to Wolf and Mintz (1957: 380) the hacienda and plantation can be seen as two similar but different kinds of social systems. A plantation they define as 'an agricultural estate, operated by dominant owners (usually organized into a corporation) and a dependent labour force, organized to supply a large scale market by means of abundant capital, in which the factors of production are employed primarily to further capital accumulation without reference to the status needs of the owners'. In comparison a hacienda has been defined as 'an agricultural estate, operated by a dominant land-owner and a dependent labour force, organized to supply a small scale market by means of scarce capital, in which the factors of production are employed not only for capital accumulation but also to support the status aspirations of the owner'.

The general conditions necessary for the development both of the hacienda and the plantation include (a) technology adequate for the production of a surplus; (b) class stratification needed to permit differential access to the factors of production and distribution; (c) production for a market, usually an outside market; (d) the injection of initial foreign capital and possibilities for capital accumulation; and (e) the politico-legal system which supports their operations.

Whilst the plantation needs a large-scale supply of capital, the hacienda usually operates within a situation of capital scarcity. One reason for this is that haciendas have limited demand on smaller markets as compared to the large-scale markets of plantations which are often supranational in scope.

It is difficult therefore to have large-scale capital generation in the small-scale arena of hacienda production. Haciendas often obtain their money from local merchant groups or small banks, which do not have the financial status necessary for large, high risk loans. Plantations obtain capital investment from corporation type entities which gauge the level of their investment on the basis of maximum returns on capital advanced. Initially, plantations need a large investment to cover their extensive fixed costs, so loans from these foreign investors are critical. As plantations are usually to be found in so-called 'Third World' countries, capital is imported and ownership is often of necessity foreign. An important consequence of this foreign ownership is the external influence on the social, political and economic spheres of the host country. There is a well-known connection between imperialism and plantation economy.

Haciendas, on the other hand, because of their low input of capital are usually within the financial means of a local person of a particular social standing who has a sound economic base. The most typical form of ownership of such estates therefore, is usually family ownership. The owner of a hacienda tends to have certain aspirations of power and prestige for his family. As a result, the factors of production in sharp contrast to the plantation type are not always manipulated for maximum profit with no reference to the status/consumption needs of the owners.

Land is another important condition for the creation of haciendas as well as plantations. They both require large amounts of space but their use of it tends to differ. Haciendas need land to grow their cash produce and to provide workers with subsistence plots and other benefits like wood or forest resources. These extra allowances to the workers are in lieu of proper wages; salaries are limited on haci-endas due to scarce capital investment. Sometimes these features are also found on the plantations, for example, the Assam tea gardens provide such benefits to their labour to compensate for lower wages.

The plantation in its pure form requires land in sufficient quantity to make maximum profits, and technology which includes not only machinery, access to engineering and other technical skills but also the means to build transport and other communication facilities. Besides these initiating conditions, sanctions of a politico-legal kind are required to give legitimacy to the planters' use of military type security arrangements for controlling the labour force (Mintz 1957: 43).

A large labour force is needed at certain times during the cycle of growth of the cash crops of both haciendas and plantations. Drawing labour from the subsistence sector for casual employment and creating a subsistence sector where it did not exist previously, were the ways devised by planters to meet seasonal demand for labour on their plantations. Very much like the migrant labour in South Africa and the United States (Burawoy 1980), plantation labour survives by depending in part on the non-capitalist subsistence economy as its own cash wages in the plantation sector are generally below subsistence level. For example, in Assam, workers are allotted minuscule plots of paddy land which provide barely 30 per cent of a labouring family's rice needs. This kind of 'largesse' from the employers is seen as 'something is better than nothing'. In Assam, with the colonial government's support, planters were able to settle ex-tea garden labourers on government land adjacent to tea gardens. At present, the planters continue to benefit from casual labour from these settlements, where ex-tea garden workers-cum-farmers are extremely poor and are willing to take up the tea garden jobs as and when they become available. Such labour is recruited by both large company-owned tea gardens as well as proprietary gardens.

Integration of Women into the Labour Class

The entry of the various socio-cultural groups from the subsistence economies in rural areas into the world capitalist system – of which the plantation system is an off-shoot – resulted in the formation of a powerless and underpaid 'labour' class. In general, the plantation class structure had Europeans occupying managerial roles, educated colonized gentry in supervisory and clerical roles and the illiterate migrants were the labouring masses. During the colonial phase, the class formation on plantations throughout the world was marked by racial and ethnic differences; managers were always white, and supervisory staff though non-white were of totally different ethnic groups from labourers. The estate manager provided an estate grocery shop and/or a weekly market, creche, place of worship, toddy shop. These fixtures tied the workers to the plantation and created in them a psychological dependence on white managers who took an interest in labour 'welfare' with an eye to the formation of a stable workforce. Along with these features, the low wage policy in the plantation sector has continued and so have patterns of controlling and managing labour to coerce it to remain on the plantations and maintain the requisite level of productivity.

In most parts of the world, plantation labour, while remaining one of the lowest paid occupational categories, has invariably included women. Their work in the plantation sector and its links to gender and class relations can be viewed in relation to the overall interaction between the industrial and the community sub-systems of the plantation (for the latter formulation see Jain, R.K. 1970: xviii). The influence of the industrial organization on different aspects of the labouring community, such as the division of labour within the family, control of familial resources, the pattern of authority in the family, rules of residence and forms of marriage, reflects the degree of control and authority the plantation system exercises on the lives of the labourers.

Reconstructing the past of Jamaican female labour, during slavery, Lucille Mathurin-Mair has dealt with reproduction of labour in a milieu where women were used on sugar plantations 'as substitutes for animals and for machines'. Colonial socio-economic imperatives predicated the use of coercive forms of control and employment of women in primarily unskilled jobs. The pragmatism of the Jamaican colonial apparatus was backed by the racist ideology which facilitated the conversion of female labour into the chief support of estate manual labour. Phrases like 'assembly line' for labour muster (see p. 21) reflect the West Indian perspective of the rise and development of the plantation sector in the region as harbinger of industrialization. In her analysis of the struggle and resistance by women workers, Mathurin-Mair captures the spirit behind 'women's subversive and aggressive strategy directed against the might of the plantation'. Her analysis of the dynamics of female responses to plantation control demonstrates the extent to which a labour system aspiring to be a total social system could on the one hand reinforce the powerful male control of the colonial capitalistic enterprise, while simultaneously challenging both the ideology and the instruments of that enterprise. Slave women were central to that dialectical process.

The end of slavery brought about many changes. What did it mean for women on plantations? Set in the framework of historical studies of Caribbean women's experiences, Rhoda Reddock's contribution brings out gender-based discrimination in wage structures, the high rate of mortality and low capacity of reproduction among migrant Indian women workers, employed under indenture contract on sugar plantations in Trinidad and Tobago. Reddock challenges the myth of 'the Indian women's docility, tractability and "natural" acceptances of dominance' as well as the myth of Indian men migrating with their families intact to work on sugar plantations. Reddock shows that the system of 'wage labour' was in many ways akin to slavery. Throughout the period of indentureship, Indian women were mostly involved in agriculture on sugar estates as well as on cocoa and coconut estates. Based on an assumption that only able-bodied men could do heavy tasks and women and children could not, women were supposed to do light work. But, as shown by Reddock, they were involved in 'heavy work'. The planters' purpose behind creating a division between heavy and light work, was to devalue women's labour. As a result, they got the labour of two workers and paid for only one and a half. Not only this, when women received rations during pregnancy, it was constructed to be a debt which was carried forward. This meant that while repaying the debt women worked and got no wages for months.

In her discussion of social organization among Indians during the indentureship period, Reddock has brought out Indian women's struggle for autonomy. The problem of the disproportionate sex-ratio (along with six other factors listed by Reddock) need not be dismissed as was suggested by Breman and Daniel (1992: 285). It is related to the reluctance of planters to bear the cost for reproduction of a new generation of workers and for maintaining women during pregnancy as well as other non-producing family members. Notwithstanding Reddock's claim that 'for most lower class/caste men, emigration had been a means through which they were attempting to improve their class and, if possible, their caste status' (for its critique see Jain, R.K. 1986: 316), there is no doubt that once the migrants were trapped in a plantation economy, gender relationships were redefined in the new situation, much to the chagrin of the Victorian values relating to women's position in society. The ultimate incorporation of Indian women's labour into the domestic economy has been related to the economic crisis of the 1880s and 1890s, caused by increased competition in the international sugar market, and increased male violence against women.

Concerned with migrant Indian women's labour under indenture contract on plantations in Fiji, Shaista Shameem also takes a historical perspective to analyse the issue of marginality of women in capitalist production. Migrant women resisted violence and brutality from both the planters and indentured male workers by withholding their labour, carrying out physical acts of violence upon male overseers and by refusing to live with their husbands. Here is not a successful tale of emancipation of women. As wage labourers women struggled against planters and male labourers, but the discriminatory wage structure forced them to depend on men for their own and their children's subsistence. Relations between the Christian church, the colonial state and the sugar companies combined to blame the victims (women labourers) themselves for their problems.

In the post-colonial era, independent sovereign states, commonly designated as the Third World, were created as a result of decolonization. Although these states are no longer under political control of the metropole, they are to a certain extent still under their economic control. Wage labour is now the accepted means of remuneration of workers but as the chapters of this book show, variations in the inequalities of class on plantations and in patterns of gender relations result from historical, socio-political and cultural factors. Rachel Kurian has shown, 'the Tamils who worked on the plantations, were, in the Sri Lankan polity, essentially a separate ethnic group, more readily identifiable by the work they did and by their harsh and strictly disciplined existence'. Based on oral interviews and historical research, Kurian has placed the contemporary plantation setting within the dynamics of historical development. She shows that in Sri Lanka, plantation labour was incorporated into a system of hierarchy which was held together by the plantation ideology, i.e. acceptance of 'the self-contained world of the estates'. Plucking, the most labour-intensive task on the tea estate and one which is of highest importance, is still done by women. The estate management prefers to employ women for this job. Both plucking and tipping, done by women, are closely supervised by male foremen (*kangani*). Authority of the male *kangani* looms large over women workers in both 'on' and 'off' work.

In South India, Sri Lanka and Malaysia, the same person, both as recruiter and foreman, is known as *kangani* (properly *kankani*, Tamil – headman). Compared with the status of a *kangani* on the rubber estates in Malaysia or on rubber or tea estates in Sri Lanka (cf. Jain, R.K. 1970: 206; Jayaraman 1975: 57–63), the power and authority of a sardar on Assam tea gardens is less marked in its exercise over his gang of labourers. Frank Heidemann, in his study of *kangani* in Sri Lanka and Malaysia, has aptly summed up the status of a contemporary *kangani*. Heidemann (1992: 36) writes,

> Today a *kangany* is a supervisor of lower rank and his duties are similar to the work of a *subkangany*, as he is the direct supervisor of a labour gang of usually 10 to 20 persons. But the *kangany* is no longer the centre of a system; the socio-economic order of the tea plantation can no longer be called a *kangany* system.

In the old days, prominent sardar (*kangani*) were appointed by the garden manager to officiate in the estate *panchayat* (the people's assembly, now almost defunct), and to conduct *bichar* (the settlement of disputes by taking into account the evidence produced by concerned parties and their witnesses and making decisions in accordance with the customs of the people).

Kurian has argued that traditional notions concerning gender roles in Tamil society in combination with caste ideology, respect for hierarchy, sanskritization process, and the resurgence of religious practices have special significance in terms of women's position. Viewed as inferior to men, women have suffered inequality as they are burdened with double responsibilities of work in the tea fields and in the households. The process of sanskritization has meant that women of the *Sudra* sub-castes have lost the freedom they earlier had. Kurian compares the *Adi-Dravida* women to *Sudra* sub-caste women, the former being 'more open with their men' and able 'to stand up and fight for their rights' while the latter are 'less free and more subservient to their husbands'. According to Kurian, there is a correlation between the level in the caste hierarchy and the degree of reticence that women display towards their men. The higher the caste, the greater the degree of reticence that they show. Women's mental and moral subjugation is coupled with physical violence against them.

In another part of the world, Indian female plantation labour in Jamaica during the indentureship and post-indentureship periods, states Verene Shepherd, remained tied to plantations even after the abolition of indentured servitude, in 1921. Like Reddock (1985) for Trinidad and Emmer (1985) for Suriname, Shepherd too believes that Indian women, migrating to the Caribbean as indentured labourers, left their homeland in order to improve their socio-economic status. This claim is subject to debate as it assumes that indentured migrants made a rational and conscious decision 'to emancipate themselves from an illiberal, inhibiting and very hierarchical social system in India'. There are two separate issues involved here. One relates to overwhelming archival evidence of fraudulent methods employed to recruit indentured labour to plantations. (This shows that the migrants were, by and large, not in a position to make a conscious decision to emancipate themselves.) The other concerns the Caribbean view of the Indian social system vis-à-vis the relatively more open and free socio-cultural dynamics of gender-relationships found locally. Insofar as Shepherd, Reddock and Emmer try to explain the phenomenon of large numbers of single women migrating to the Caribbean and link it to the socio-cultural life of Indian immigrants in the Caribbean, the logic of their argument has to be placed in the Caribbean context. More relevant here is the fact that even though the planters did not want to employ women on estates and therefore kept their wages much lower than those of male migrant labourers, the poor economic status of Indian men caused even married men 'to privately ask planters to send their women out to work'.

Working through dense archival material Shepherd has analysed the wage-structure to show how women workers were unable to earn as much as men even when they worked more than them. Their plight in Jamaica became worse in the 1930s due to the impact of the economic depression. Jobless Indian women became destitute, left to fend for themselves. Shepherd traces the conditions of Indian women in Kingston. Largely single women, according to the population census of 1943, the majority of them legally married, some lived in common-law relationships while a small number were widows. Either engaged in cultivation and supply of flowers or in the service industries in urban areas, many of them were in low-paid jobs as domestic servants. Shepherd has argued that the gender ideology of men's place being in the public domain and that of women in their homes shaped the social and economic experience of female Indian plantation workers during and after indentureship.

Female labour in many plantation communities has been reported to be doubly exploited and oppressed by the capitalist economic system. Reflecting what is happening in the Sri Lankan society at large, Kurian (1982) in her study of its Tamil female labour and Daniel (1981) in his study of its estate Tamils provide examples of a system heavily discriminating against women. Ramaswamy (1993: 127–8) has also cited Tamil songs relating the pitiable condition of women working on farms and tea gardens. Similarly, Devaki Jain (1980) in her study of family planning among the female workers on the South Indian tea plantations found that the system subjected them to the worst forms of exploitation and oppression. In contrast, Sutton and Makiesky-Barrow (1977) demonstrate the 'absence of marked sexual inequalities' in the Afro-Caribbean situation. C. Jayawardena (1963) also described the ideology of social equality, including sexual equality, found among the sugar plantation workers in Guyana. Notwithstanding variations in analyses of discrimination and oppression of women meted out by plantation systems, the Caribbean studies point to relative lack of sexual inequality in both the Afro-Caribbean and the Indo-Caribbean labouring communities on sugar plantations. Assam tea gardens in a post-colonial setting, present neither an isolated nor a unique example of a measure of equity in gender relations, nor a miracle without a historical or contemporary reference.

Based on anthropological fieldwork on a tea garden Shobhita Jain has shown that virtually complete stoppage of fresh recruitment of labour to the Assam tea gardens coupled with several mechanisms of tying the existing labour to plantations meant the gradual formation of a stable workforce. The concept of family wages pushed women into the labour market and the planters' barely disguised hostile attitude towards family planning programmes for the labourers indicates that the 'control' of reproductive forces may not be all that unintentional. At present permanent workers live in the garden's labour 'lines' while the majority of casual workers are drawn from those ex-tea garden workers-cum-farmers living on government land adjacent to tea gardens. The tribal social structure, which pervades the labouring women on Assam tea gardens, has traditionally allowed certain 'choices' and freedom which were never translated into the 'prostitution' associated with women in Fiji, north Sumatra and the West Indies. Also, a labour policy of engaging females for the main task of plucking and male labourers for secondary tasks of weeding and of levelling down the economic position of sardar (always a male) with other workers, has taken away the edge of superiority on the basis of gender. The degree of parity in gender relations that is obtained on the tea gardens is however not an indicator of general well-being of the workers as a class.

No case in this collection has shown that plantations have provided fair wages to the workers. Sugar-cane workers' households on haciendas in Negros Island of the Philippines, studied by Violeta Lopez-Gonzaga, repeat the same story. With low earnings, large family size, low levels of education, when men are rendered jobless in the wake of a slump in the island's sugar-cane industry, it is the Negrenese women who ensure their family's survival by taking up sundry jobs and petty forms of entrepreneurship. Women working on sugar-cane plantations are given the tasks of weeding and fertilizing. This is in keeping with the stereotyped gender-based allocation of work to them. They are not given maternity leave benefits. When engaged in economic activities outside the plantations, they show a great deal of resilience amidst absolute poverty. In some ways, the Filipinas of Negros Island and the Assam tea garden women workers share the ability to secure survival amidst excessive stress. Both sets of women show a high degree of flexibility and adaptability which are manifest in articulating the household division of labour, contribution of children to housekeeping, financial management and cost control, exchange of goods, resources and services on the basis of one's familial network.

Piet Konings' contribution reflects the process of integration of women in the African plantation sector where, as mentioned earlier, women workers are nearing 50 per cent of the plantation labour force. Even though the tea estates in Anglophone Cameroon managed to recruit female labour, Konings argues that the management found it difficult to control the labour process involving the working women who protected their interests both individually and collectively. Konings' chapter seems to be the logical conclusion of this collection of case-studies of plantation women emerging as active and aware agents of action. We do not know if the so-called patriarchal control of African, Indian, Indonesian and Philippine women has really been objectifying them but the authors in this volume appear to claim that they have begun to object. Apparently, no matter how much the subject subjects the 'other', the 'object' simply objects and is no longer willing to be treated as a commodity.

The multi-disciplinary approaches used in this collection bring out cross-cultural perspectives of female labour in the plantation sector and show that both the value and fact of women's productive and reproductive work are largely mediated by institutionalized patterns of male authority and domination. Yet, the range of variation in women's position in the plantation milieu indicates that male dominated family life is slowly changing as authority is shifting from male heads to conjugal pair, from family head to earning members of the household. All the same not a single case-study has indicated that there is any perceptible change in the exploitation of labour. The class and status inequalities in the plantation sector remain its hallmark and to that extent capitalist (read male dominated in the context of gender relations) control of labour, including female labour, negates the significance of work as an index of higher status or prestige. This book has presented both the gloomy images of women's subjugation and some fissure in patriarchy. Undoubtedly, much further exploration is required after this modest beginning.

2

Women Field Workers in Jamaica During Slavery*

Lucille Mathurin-Mair

Any attempt to reconstruct the past of Jamaican female agricultural labour is bedevilled by the colonial or metropolitan orientation of much of Caribbean historiography. This is perhaps inevitable, given the primacy of European strategic and commercial interests in the establishment of New World plantations. A direct consequence is that the enslaved African men and women who laboured in those plantations have been for centuries submerged in the archives, barely making it to the footnotes. Women have suffered as well from the invisibility which has been the nearly universal fate of women's work, whether slave or free, in or out of the field, in the past or the present. When noticed, it is usually seen as marginal to the national, not to mention the international economy. This applies to that wide range of essential goods and services which women produce in the home; they are seldom if ever quantified, seldom if ever regarded as significant enough to be reflected in calculations of national income. This is equally true of women's enterprises in that sector of the economy, which, in Jamaica and the rest of the Caribbean, is dominated by women, and is variously labelled the informal or parallel or underground sector, all of which imply activities that are imprecise, irregular and out-of-sight. This syndrome of virtual non-existence constricts research. Economists who, for example, wish to compute female rural labour in regions of Africa, Latin America or Asia, find their attempts to get hard sex-aggregated data frustrated by a widespread insistence on the part of men, but of women too, that women do not work; and this, despite the clear evidence of the vital agricultural tasks in which women have been engaged for centuries.

For the Caribbean region, the 'new' history of slavery, in which Goveia's (1965) work was a major landmark, allowed us to adjust that distortion of female reality. In contrast to a generation ago, the greater body of knowledge at our disposal today about that formative period of Caribbean societies, gives us the ammunition to reject the stereotype of a female labour force which is peripheral to the dominant

---

* Elsa Goveia Memorial Lecture, The University of the West Indies, Department of History, 1986.

sectors of the economy. Scholarship, such as Patterson's (1967) work on the sociology of Jamaican slavery, Craton's and Walvin's (1970) study of Worthy Park estate, my own research findings (Mathurin 1974), and Higman's (1976) meticulous demographic analysis of the Jamaican slave population on the eve of Emancipation, all testify to the extensive involvement of slave women in Jamaica's agricultural enterprises, and notably the prime productive processes of the sugar industry.

It was sugar which placed Jamaica at a strategic point in the emerging international capitalist system of the eighteenth century, establishing it as Britain's most prized transatlantic colony. In 1805 it was the world's largest individual exporter of sugar. Sugar commanded the island's major resources of land, capital and labour. In 1832, sugar employed 49.5 per cent of the slave force. The majority of those workers were women, the ratio being 920 males to 1,000 females (Higman 1976). It was the requirements of sugar that set the occupational norms for the bulk of the population, even for those workers not directly involved in sugar production. And it was those requirements, as identified by the captains of the sugar industry, that dictated a conscious policy of job allocation which concentrated black enslaved women in the fields in the most menial and least versatile areas of cultivation in excess of men, and in excess of all persons, male and female, who were not black. Patterson (1967) examined the distribution of the slave labour force on Orange River and Green Park estates in the parish of Trelawny, in 1832, as well as on Rose Hall estate in St. James in the 1830s. He was struck by the large numbers of women who were engaged in field tasks as contrasted with the men, who were spread over a wider range of skilled, non-predial jobs. He concluded that the most distinctive occupational feature of Rose Hall estate was the preponderance of field women – one to every two – as compared with one man in every eight. He attributed this partly to the fact that men had more job options available to them, and partly to the fact that women out-numbered men in the Rose Hall workforce. This indeed was typical of the demographic situation throughout the island after the abolition of the slave trade in 1807 when the population structure was no longer manipulated by human importation, and moved by natural increase from a male to a female excess. However, there is considerable evidence (see Bush 1990: 33–50; Shepherd et al. 1995: 125–72) to indicate that women were tilling Jamaica's valleys for nearly a century before that period, and had been performing the most arduous tasks of agricultural cultivation, in greater numbers than men, even at an earlier time in the island's development when men outnumbered women in the slave population.

The pattern of sex-differentiated labour deployment evolved out of colonial socio-economic imperatives as well as out of traditional concepts about the sexual division of labour. In the seventeenth century, Jamaica was largely an island of small proprietors engaged in hunting, and in subsistence planting, with some cash crops such as tobacco. During this stage of relative underdevelopment, estate work was structured roughly along the following lines: a class of white indentured servants did some field work, but for the most part supervised the slaves and carried out the skilled and technical functions of the estate. This servile white group numbered approximately 20 to 30 per cent of the labour force. One of the rare censuses of the period, for example, in the parish of St. John in 1680, listed 95 white servants and 527 slaves (Bennett and Heylar 1964). Male slaves bore the brunt of the more physically demanding field labour. Female slaves, who at that period were present in approximately equal numbers to male slaves, supplemented male labour in the lighter tasks of the field, and worked as domestics in the planters' houses. White indentured servants had the advantage of race and of short-term bondage, and these enabled many to climb the creole socio-economic ladder, eventually to become property owners, or to migrate. As they moved up or out, taking with them their skills, male slaves were recruited into those areas of greater expertise which whites were vacating. The colonial establishment only grudgingly conceded this process and the local assembly, by a series of legislative measures in the early eighteenth century attempted to bar male slaves from supervisory positions, from trades, and from the more expert technologies of the plantation (see CSP 1716–17). From this time we see Jamaica's transition to a highly capitalized monocultural export economy operating in large plantations. This process accelerated the flight of small proprietors as well as of the white indentured class. Male slaves continued to fill those work slots originally conceived for lesser whites. The occupational vacuum which they, in turn, left in the fields, was filled by women.

Estate inventories of the 1720s and 1760s, which have been examined, make clear the wide range of specializations available to male slaves. They also indicate the limited choices which women had (see JA IL 1720–71: 14, 44, 46). Particularly rich archival material exists in the estate papers of the absentee planter, William Beckford, who was also the Lord Mayor of London in the 1780s. He owned a complex of twelve properties, spread throughout the parishes of Clarendon, Westmorland and St. Ann. The majority produced sugar; the others were livestock and cattle pens. Together they provide revealing evidence of the conscious deployment of a labour force of 2,204 slaves containing a slight excess of men – 802 males and 778 females (Beckford 1773–84:C.107/43). A minority of the men, viz., 291 or 36 per cent of the male total were field workers, compared with a majority of the women, 444 or 57 per cent of the female total. The men who were not field labourers were among other things, stonemasons, blacksmiths, fishermen, farriers, wheelwrights, wharfmen, coopers, sawyers, doctor's assistants, tailors, distillers, bricklayers. Many of these jobs involved the kind of special expertise which placed such men among the elite of the slave hierarchy. They sometimes combined predial jobs with skilled occupations, so that a field man might also be a boiler in a sugar factory. In contrast, women were confined within a much more restricted area, and field women were exclusively field women. Where they were listed in other categories of work apart from that of the field, a close look shows that they were often engaged in ancillary field tasks such as cutting grass.

Domestic work was the most significant work category next to agriculture, and accounted for 59 or 13 per cent of the sample. Within this group were washerwomen, housewomen and cooks. A significant number of house slaves, nine out of eighteen, were mulattos, consistent with the creole view of coloureds[1] as being incapable of physically demanding labour, such as that of the field. Elsa Goveia's observation on the occupational rating of domestics is here relevant: the female house slave on the plantation, she noted, was not so much a 'skilled' as a 'favoured' worker. Women with acknowledged 'skills' numbered 34 or 8 per cent of the Beckford slaves, and included midwives, doctresses, field nurses and seamstresses. But even this group of non-predial female workers overrates the black women's access to special expertise. Mulattos, for example, dominated as seamstresses – eleven out of a total of fourteen; and the labels of doctress, midwife and nurse were often euphemisms for superannuated field workers, regarded as physically unable to continue in arduous agricultural work, but just about capable of providing services for younger workers. The period of the 1780s to which these data relate, was the period when the traffic in African slaves was at its peak, and planters had a clear policy of sex-selective importation of labour with a preference for men. As a consequence, African men constituted the majority of the workers on the plantation, outnumbering women right until the time when the slave trade ended. It was in the context of this sex ratio of male excess in the workforce that a pattern of job allocation evolved in which not only were the majority of the African women field labourers, but the majority of field labourers were African women. And this had profound implications for their working lives.

The routine of plantation agriculture with its para-military regimentation is widely documented (see Roughly 1823: 55–71). Youth and physical fitness were the main criteria by which manual tasks were distributed. The first or great gang of the workforce, 'the flower of all the field battalions', comprised both sexes, from ages fifteen to about fifty years, and was 'employed in the most laborious work'. According to Cooper (1824), 'in the extraction of this labour, no distinction is made between men and women' on a sugar estate, that involved digging cane holes, planting, trashing, cutting, tying, loading and carting. Female labourers used the same implements as male, the bill and the hoe. The most strenuous work in the agro-industrial cycle of sugar production operation took place during crop-time when continuous supplies of freshly cut stalks of ripe cane had to be processed without delay. Workers were then occupied from sunrise to sunset, averaging six days and three nights each week. A witness to the British Parliamentary Commission of Inquiry into the state of West Indian Slavery in 1832 (see *Parliamentary Papers* 1832: I, 127) testified as follows:

Q. Are the women employed for the same number of hours as the men?
A. Yes, except the women who have children at the breast.
Q. And at the same description of labour as the man?
A. Almost entirely, there are of course different branches of labour which they cannot undertake; they cannot undertake the management of cattle; they are excused from night work out of crop, in watching. (That is the women do not guard the crop at night).
Q. Are they employed in digging cane holes?
A. Yes.
Q. In gangs with men?
A. Yes.
Q. And exposed to the same degree of labour?
A. Yes.

It is worth noting that as Higman's (1976) data reveal, women spent more of their working years in the field than did men.

It is interesting to consider the two prevalent gender assumptions of the period regarding women's physical and intellectual capacities, assumptions which can be bluntly stated as follows: Women are not strong enough to do tough manual work, nor are they clever enough to master things technical – both assumptions well proven to be without foundations, but both frequently invoked as strategies of female subordination. The Jamaican colonial establishment was, however, distinguished among other things for its pragmatism, which was unfailingly inspired by economic self-interest which enabled planters, without reservation, to discard the image of woman as a frail creature, and convert her into the mainstay of estate manual labour. At the core of that mental flip-flop was the racist ideology which supported African enslavement and which denied Africans normal criteria for judging human potential. Racism eased the process by which black women, and black women only, were massively drafted into the 'assembly line' of the field.

The other gender assumption concerning women's technical capabilities proved to be of hardier stock. It was bad enough that planters had to overcome their aversion to the idea of a male slave with expertise, but to be female and skilled was unthinkable. And this had wider consequences for the political economy of the plantation system, functioning as it did in a vicious circle of labour exploitation and retarded technology. The Jamaican economist, Girvan (1976: 158), has analysed this phenomenon with great insight; and in so doing, has placed today's Caribbean technology policies in a relevant historical continuum. He writes, 'If the artisan workshops of Europe were the basis of technological innovation, the mines and plantations of America were a prescription for technological stagnation . . . The slave plantation system systematically underdeveloped the technical capacities of the population.' A gender dimension to that analysis even further sharpens its relevance. Little scope existed in Jamaica's mono-crop economy for the mixed farming enterprises, dairy production, and cottage craft industries which rural people in other economies developed; and of all labourers on the plantation, it was the black female who was the least able to diversify or upgrade her economic potential. She could not, for instance, perpetuate the considerable range of skills and crafts characteristic of the African rural economy in which she had been a central productive figure before she crossed the Middle Passage.

Furthermore, female labour on the estate account book was expensive labour because of the inevitably high rates of absenteeism associated with child bearing and child rearing. So unless she was producing children and providing additional work units in numbers sufficient to compensate for her absences from the field gangs, the female slave offered relatively minimal returns on capital outlay. To offset the cost of her maintenance, she had to be heavily utilized, and this need to extract maximum manual output from female labour was a factor in the plantation's reluctance to rationalize its operations. The planter/novelist Matthew 'Monk' Lewis (1834) tried to promote 'the labour of oxen for that of Negroes, whenever it can possibly be done'. The creole historian/proprietor Long (1774) estimated that 'one plough could do the work of a hundred slaves and should be encouraged . . . for no other work on a plantation is so severe and so detrimental as that of holding or turning up the ground in trenches with hoes'. But both the economics and the politics of slavery mitigated against the introduction of oxen and ploughs. Estate managers and overseers, for example, had vested interests in their own slave property, whom they hired profitably to the estates for miscellaneous jobs such as holing, hoeing and planting. They were a significant pressure group in the creole resistance to technological progress. Moreover, the power-relationship inherent in a coercive labour system obscured considerations of managerial efficiency. The plantation establishment saw the slave as a special kind of multi-purpose work equipment, a flexible capital asset which, unlike animals or machinery, could be deployed for almost any use at any time. Women were even more multi-functional, for, of course, they could also replenish the labour supply. So as long as the most technically backward, but versatile section of the labour force, viz. slave women, serviced the plantation in large and growing numbers, little incentive to technological progress existed. And as long as Jamaican agriculture remained technologically retarded, women continued to be used in the field in overwhelming numbers as substitutes for animals and for machines.

And what did this mean for women's status? Elsa Goveia (1965: 233) wrote of the 'ordinary field slave' that

> no other group of slaves was so completely subject to the harsh necessities of slavery as an industrial system . . . they lived and worked under the discipline of the whip . . . they had fewer opportunities for earning a cash income than most other slaves . . . they were

maintained on the bare margin of subsistence . . . although they did the most laborious work, their standard of living was severely lower than that of any other group of slaves.

An important index of status was the monetary value of the slave. The highest figure attached to a female slave was normally that of 'a strong, able field worker'. For the man, that tended to be his lowest price, for his market value rose as he acquired a skill. An official valuation of 1789 showed the most expensive women, field women, averaging between £75 and 85, with the exception of one midwife who was worth between £150 and 200. Male predial workers were slightly more costly than their female equivalents, being priced at between £80 and 100. Eleven different categories of skilled male workers were also listed. Their prices started at £120 and rose to £300 (see Beckford 1783–84). The market had a low assessment of field labour which was regularly used as a punishment. No greater disgrace could fall to the lot of an artisan or a house slave than to be demoted to the field for some misdemeanour. This implied, more often than not, besides an instant decline in status, physical indignities imposed 'under the discipline of the whip'.

Not surprisingly, field women's health was generally poor. Thirteen per cent or 188 of the 604 women and girls found on the Beckford estates were in various stages of physical disability. Of 199 who were over the age of forty, 68 per cent suffered from impaired health. The condition of the young female adults was particularly revealing because they, presumably, were in the fittest condition to perform the demanding field tasks allocated to them. In fact, 22 per cent of that prime group were listed as in poor shape. Few diseases or illnesses as such were diagnosed, but these young women were variously described as 'weakly', 'infirm', 'distempered', and so on, suggesting a general debility associated with inadequate diet, inadequate care, excessive exertion and physical abuse. And this had grave consequences for, among other things, women's childbearing functions. Beckford's Clarendon estate, with a population of 274 women of childbearing age, produced nineteen live infants in 1780, a year in which slave deaths numbered thirty-four. The decrease of the slave population was in fact a Jamaica-wide phenomenon estimated to be approximately 2 per cent per annum in the 1770s and 1780s.

It was within this depressing demographic context that the abolitionist and/or humanitarian movement gained momentum, and generated pro-natalist measures, starting with the Consolidated Slave Acts of 1792 which gave official cognizance, for the first time, to the role of the female in population growth. Clause XXXIV of such legislation, for example, provided that every woman with six children living should be 'exempted from hard labour in the field or otherwise' (Edwards 1801). The abolition of the slave trade in 1807 saw such policies intensified. The legislation of the period, while containing some elements of concern for the well-being of slaves, was primarily motivated by the need to sustain labour supplies. Various rewards were devised for mothers, 'adopted' mothers and midwives who had healthy children. An example of the favoured status which prolific women received during the last decades of slavery is found on Matthew Lewis' estate at Canaan in the parish of Westmorland. He instituted an order of honour for mothers, which gave them the privilege of wearing a 'scarlet girdle with a silver medal in the centre' and entitled them to 'marks of peculiar respect and attention'. He even declared, for the whole estate, a 'play day' at which the mothers were special guests. He made sure that all the slaves knew they owed their good fortune of a holiday to the 'piccaninny mothers; that is, for the women who had children living'. Between 1816 and 1817, however, at about the same time that Lewis was celebrating motherhood on Canaan estate, the slave population had reached its numerical peak of 346,150 and moved steadily into a decline which only ended with emancipation (see Roberts 1957).

One is perhaps here observing an example of that well-known propensity of colonial designs to self-destruct. The Jamaican establishment, while ostensibly assuming social responsibility for reproducing the labour force, was simultaneously pursuing economic policies virtually guaranteed to diminish that labour force; for official urgings and incentives to women to be fruitful and to multiply went hand in hand with the oppressive work loads of the field, and punitive practices which had profound impact on women's ability and desire to bear and rear children. The highest mortality and the lowest fertility rates were found on the sugar estates, establishing a strong correlation between the labour requirements of the island's main industry and the decrease of the slave population. Slave women also themselves exercised effective control over their reproduction. They practised abortion, they prolonged lactation, they drew extensively on the midwives' knowledge of birth processes in order deliberately to depress their fertility. Both voluntarily and involuntarily, they frustrated the establishment's hopes for a self-reproducing labour force. They could not escape the backlash of that frustration.

Creole attitudes towards women noticeably hardened in the latter years of slavery, almost in direct proportion to the plantocracy's dependence on women as a key source of their labour power. The most highly publicized cases of planter brutality are recorded in the 1830s. In most instances the victims were female slaves (see *Parliamentary Papers* 1832: II, 127). A telling example of the heightened antagonism towards women is the Jamaican Assembly's refusal to regulate or to abolish the flogging of women, despite repeated recommendations for reform from the colonial administration in Westminster. Not even for aged or pregnant women would the plantocracy consider exemption from corporal punishment. Jamaica exactly reflected the view of the Barbadian Council which debated the issue at the time, and which declared that to abolish the flogging of female slaves 'would mean adieu to all peace and comfort on plantations' (Williams 1952).

Women became even more conspicuous targets of planter hostility during the apprenticeship period which was mandated under the Abolition Bill of 1834 to phase out slavery in gradual stages. Ex-slaves, now apprentices, were required to give 402 hours of work weekly on the estates to which they were attached. The rest of their time was theirs to place and barter on the wage market, if they chose. This stage of quasi-bondage was originally conceived as a six-year system. It proved to be 'an impossible compromise' and was terminated after four years. For the period that it lasted, planters used apprenticeship as a last stand to retain maximum control over what they saw as a dwindling labour force (see Hall 1953). The increasing withdrawal of the female apprentice from the cane field was a critical factor in that perception. One contemporary (see BM 1838) analysed it aptly as 'women's rejection of that type of labour which is inseparably associated in their minds with the idea of torture, oppression and degradation'.

Thomas McNeil, attorney for the estate of Lord Holland in Western Jamaica (see BM 1838), expressed the plantocracy's obsessive fear of a labour shortage and the need for an adequate supply of female workers to avert such disaster. He calculated that even if the men gave sustained labour it would still be quite impossible to continue sugar cultivation to give any remuneration to the proprietors 'unless the females also be induced to labour regularly'. 'Inducement' of women took many forms, and was facilitated by the Abolition Act itself, which entrusted to the planter-dominated local legislature the detailed subordinate arrangements necessary for the implementation of the law. Important aspects of the apprentice's welfare fell to its discretion; and, despite the appointment of special stipendiary magistrates to mediate between masters and apprentices, abuses and injustices were rampant and bore heavily on women. The Select Committee of the House of Commons appointed to inquire into the working of the Apprenticeship System reported, in 1837 (see Hatchard 1837) that women were 'the principal sufferers' from various measures devised to manipulate and coerce the labour force.

The maternal allowances and privileges which had been customary before 1834 fell into a grey area of the law, and were frequently either curtailed or abolished. Women were being called on to work in the field in advanced stages of pregnancy, whereas they had previously been entitled to exemption. One stipendiary magistrate maintained that on several occasions when he attempted to protect such women from excessive labour he was threatened with action for damages by the estate owners involved. Mothers with over six children found their pre-emancipation right of work exemption reversed and they were frequently sent back to the field. The services for field workers which were provided by cooks, water carriers, midwives and nurses were often withdrawn, and the women providing such services were redeployed to the field gangs.

Planters feared that apprentices, given an opportunity, would seek and find other occupational options, outside of their obligatory 402 hours estate time. And they had good reason to fear. For during slavery, women, despite overwhelming odds, had been engaging in semi-independent economic activities which offered prospects for a viable peasantry in a free society. It was the plantation system itself which inadvertently gave them the main resource for such initiatives, for the priority which the macro-economic system placed on cash crops for export, to the exclusion of food crops for local consumption, resulted in slaves having to assume responsibility for feeding themselves. Female slaves, as a consequence, like male slaves, enjoyed customary, and after the 1792 slave legislation, legal entitlement to their provision grounds. And there they put to their own and their family's use the chief expertise which the economy allowed them. Women laboured hard and profitably in their own fields. They thereby generated a historical process which was to give Jamaican women, as food producers and distributors, a pivotal role in the domestic economy.

The diary of a St. Ann doctor/planter, James Archer, reveals (see BM 1838) how threatening such slave entrepreneurship appeared to estate owners. Archer imposed strict taboos on the type of crops which apprentices could plant on their ground allotments. He also ordered that 'no additional help should be used for cultivating these plots. The person, male or female, who has any sanction to work such land must work the same himself, or herself; no brother, sister or any other person to assist.' This created special hardships for women accustomed, as they frequently were, to pooling land and labour in a family enterprise. Women provoked such spiteful measures from the establishment, not only because they could divert their time and energy to their own purposes and could so undermine the estate's labour power, but also because they could withhold their children's labour; for according to the Abolition Act, children under six years of age were to be free. As they became of age to be apprenticed, they could be employed with their mother's consent.

Planters never lost sight of young blacks as potential estate workers. But women stood in the way of their designs. As a contemporary observer expressed it in 1835, 'A greater insult could not be offered to a mother than by asking her free child to work.' At Silver Hill in St. Andrew, an overseer asked a woman to let her eight year-old son do light work on the estate in return for his clothing and an allowance. She flatly refused (see Sturge and Harvey 1838). The Select Committee of the House of Commons on the workings of the apprenticeship system reported (see COF 1835–1917: 137/2) the evidence of one witness, that Negro mothers have been known to say, pressing their child to their bosoms, 'we would rather see them die, than become apprentices'. Jamaica's fractious females became the subject of extensive official dispatches. They were singled out for their 'lack of co-operation, their ingratitude, and their insulting conduct. They were on all occasions, the most clamorous, the most troublesome and insubordinate, and least respectful to all authority. None of their freed children have they in any instance apprenticed to their former masters.' Women unequivocally stood firm on this issue, and of the 39,013 slave children who were less than six years of age on 1 August 1834, only nine were released by their mothers for estate work during the four years of apprenticeship (Hall 1953).

Motherhood, with its biological and customary social implications, is frequently perceived as a conservative force which imposes constraints on female activism. It became, however, in this instance, a catalyst for much of women's subversive and aggressive strategy directed against the might of the plantation. When they could, they withheld their labour and that of their children from the dominant socio-economic system. By their actions during slavery and apprenticeship, they placed themselves in the very eye of the storm of Jamaica's post-emancipation crisis. It was the crisis of land and labour, which reached a flashpoint in the Morant Bay Rebellion of 1865, and which ultimately created the jobless, landless proletariat of today – a proletariat which is largely urban and largely female, and whose significance in the evolution of Jamaica's modern labour movement is only now beginning to be illuminated by feminist scholarship (French and Ford-Smith 1986).

Some measure of the agony of free labour in nineteenth-century Jamaica was manifested in the exodus of the black population in large numbers from estate labour, a partly voluntary and partly involuntary process. In 1844, the year of the first official census, 80 per cent of the island's workforce was engaged in agriculture. A hundred years later, the proportion was 47 per cent, and it continues to fall. Female participation in agriculture fell from 57 per cent in 1921 to 28 per cent in 1943. In the 1980s, it averaged 15 per cent (SYJ 1981). As women attempted to escape the 'assembly line' of the field, they moved into the peasant domestic food economy, and increasingly into the urban service economy, in search of new options for economic viability and dignity. They encountered forces in the free society which would deny them these options. And so their search continues.

Note

1. The view that coloureds should be engaged in non-agriculture was a widely held view in local planter circles (see Higman 1976; Roughly 1823).

3

The Indentureship Experience: Indian Women in Trinidad and Tobago 1845–1917

Rhoda Reddock

Introduction

In addition to its predominantly African-originated population, the Caribbean region has sizeable numbers of people of Indian origin (popularly referred to as East Indians). Popular generalizations about Caribbean women, often incorrect in relation to Afro-Caribbean women, fail to take into account the specific experiences of other groups of women in the region. In those Caribbean countries where sizeable Indian populations are present, equally unfounded myths continue to define the Indian woman's experience and to differentiate her from her African sister. Yet in the wake of the development of the Caribbean Women's Movement, and in light of empirical and historical studies of the Caribbean woman's experience, the Indian woman's experience has been relatively ignored. This study, like all ground-breaking work in women's history, has the task of challenging tenaciously-held myths which, like all myths, have been useful in establishing and maintaining dominance and control. Myths about the Indian women's docility, tractability and 'natural' acceptance of dominance, were tied up with the popularly held view that the indentured Indian, unlike his African slave predecessor, had been allowed to bring 'his family' intact from India and maintain it throughout indentureship.

The research into the indentureship experience of Indian women is but the start of an area of scholarship which could greatly increase our understanding of the complex reality of the Caribbean. More specifically, this chapter will examine the work experience of Indian women during the indentureship period, as well as the relationships between women and men within the context of this colonial, plantation economy. It will also illustrate the ways in which relationships between the sexes are not merely 'private' but in fact overlap into 'public' domains and affect local and international economic and political policies. I bring out individual Indian women's relative autonomy and the constraints brought about by the dynamic of social relations on the sugar-cane plantations of Trindad and Tobago.

The Historical Context

In the wake of the abolition of slavery in the British Caribbean in 1838, the immediate post-emancipation period saw major changes taking place in the social relations of production on the plantations. One of the most important developments of this period was the large-scale introduction of Indians as bonded or indentured labour for the estates. It has been argued that, from its very inception, the introduction of Indian labourers was based more on the planter's desire to depress wages and re-assert his control over ex-slave labour, than on any apparent shortage of labour. Cumpston (1953: 72), for example, noted that in the mid-nineteenth century, Burnley, Trinidad's greatest advocate of Indian indentureship, pointed out in its support, that the introduction of 8,000 African immigrants had not been sufficient to reduce wages in Trinidad, thus necessitating even larger numbers of immigrants. Similarly, C. Kondapi (1951: 16) noted that even in territories where ex-slave labour was available, Indian labour was used to depress wages.

Indentureship, therefore, was an attempt by the plantocracy to re-establish some of the power and control which they had during slavery. Labourers were contracted or indentured to a particular estate for a fixed period of time for fixed wages. Large numbers of Indian migrants were taken to British Guiana, Suriname, Jamaica and, to a lesser extent, the Lesser and Greater Antilles, as well as to Trinidad and Tobago. This system was one in which migrant labourers were indentured or 'bonded' to work with plantations for a fixed period of time, initially a series of one-year contracts, but later, five years. In many ways, this system of 'wage-labour' was akin to slavery. Movement off the plantation was severely curtailed. Tickets of Industrial Residence similar to 'passes' were needed. Absenteeism was punishable by fines and/or imprisonment.

Indian indentured migration to Trinidad began in 1845 when the *Fatel Rozack* arrived with two hundred and six men and twenty-one women on 30 May (Cumpston 1953: 99). From the inception, therefore, what later could be described as the 'Indian Women Problem' had already reared its head (DPRO 1915–18: 571/1–5). Recruitment was organized through two Emigration Agents, in Calcutta and Madras, who worked through sub-agents to 'entice' or 'encourage' people to migrate. The majority of migrants represented the landless unemployed – a group which had emerged as a result of destructive colonial policies which had annihilated the Indian cotton industry and, to the detriment of smallholders, had concentrated land in the hands of a class of landowners. This situation was compounded by the effects of recurrent Indian famines in the nineteenth century, which affected peasants and rural artisans (Jha, not dated II: 3).

As far as the recruitment of women was concerned, available records (as well as oral history evidence just coming to hand) suggest that additional factors were at play. These included the need to flee circumstances such as the ban on widow re-marriage, the problems of pregnancy outside of wedlock and difficulties within their domestic situation. Within the Trinidad and Tobago situation, every effort was made to keep the African ex-slaves away from the Indian migrant labourers, both socially and geographically, originally with the aim of maintaining the latter as a labouring population available primarily for plantation and later for peasant production for export. As time went on, the positioning of Indians in the colonial society included defining Indian women in terms of their difference and 'otherness'. This served to perpetuate the suspicion between the two major racial groupings as well as the Indian man's desire to 'protect' his women, already in short supply, from the women of 'the enemy'. The reclaiming of the Caribbean women's history of work in production and daily life is a necessary weapon against continuing attempts to encapsulate them within a tradition of 'housewifery' or dependence which was never theirs. This chapter seeks to illustrate this as well as to identify the precursors of this domestication process.

Women's Labour during Indentureship

The Sexual Division of Labour In Plantation Production

Before entering into a detailed discussion of the work experiences of Indian women during indentureship, it might be worthwhile to refer to the premise from which this study begins. The premise that Indian and African women at the point of their entrance into Trinidad and Tobago society were workers and not wives, is stressed once more in this section. The 1891 Census of the Indian population, commented on by government Statistician, H.J. Clarke, quoted in D.W.D. Comins' Report (1893), showed that the total adult Indian population was classified as being involved in the occupations shown in Table 3.1.

Throughout the indentureship period therefore, a vast majority of Indian women in Trinidad and Tobago were for the most part involved in agriculture, predominantly on sugar estates, but also on cocoa and coconut estates. Labour on these estates was organized on a task basis. On an average estate, tasks existed

Table 3.1. Adult Indian Population Classified by Occupations

Occupations	Total	Per Cent	Male	Female
Agriculture	40,902	78.33	26,771	14,131
Official or Professional	216	0.42		
Domestic	1,242	2.38		
Commercial	861	1.65		
Industrial	7,013	13.6		
Unoccupied (in household work)	1,891	3.62	804	1,081

Young Indian Woman during Indentureship in Trinidad. Trinidad and Tobago late nineteenth–early twentieth century

for drivers, cart drivers or watchmen. There were also tasks in weeding, forking (flat), ploughing, furrowing, banking, manuring (pen and imported), cane planting, cane cutting, burying trash, stocking, stock-keeping, supplying, truck-loading, carrying fuel and doing mill work. In any one week, five tasks of nine hours each had to be completed in the field, while in the factory, tasks could last up to fifteen hours. During crop time, fieldwork could be extended to six nine-hour days a week (Brereton 1974: 29–30; Comins 1893: 3).

Within the estate, an established sexual division of labour existed. This was based largely on the assumption that able-bodied men could do full tasks of heavy work, while women, weakly men and children could not do full tasks of light work. In addition, during the earlier period, the highly skilled jobs were done almost totally by men, mainly African ex-slaves; but later Indian men became involved. Women, therefore, did mainly weeding, manuring, supplying and cane-cutting, and these, though necessary tasks, were the lowest paid occupations. Most able-bodied male labour was concentrated on forking, cane-cutting, truck-loading and mill-work. Draining and banking, though field labour, was done mainly by free, specially skilled Indian or African labour. All other skilled tasks were carried out by men. Small children were employed mainly in grass collecting or manuring, using imported manure. According to various reports, this sexual division of labour was not constant. Major Comins (1893: 3), in his report, found that 'many coolie (Indian) women work alongside the men and do full tasks in heavy work such as loading canes in carts and trucks'. Similarly, at Esperanza and Phoenix Park, women did both light work and heavy tasks; while at Brechin Castle Estate he saw women, girls and boys doing manuring and supplying. Truck-loading was officially a heavy male task, but apparently many women preferred doing this kind of work.

Having been asked to report to the Indian government at a period when the indentureship system was being attacked, Comins repeatedly contradicted himself on this question. At the same time that he reported on the work activities of the women, he also had to show that this work was not compulsory, and that the possibility existed to withdraw women (although only free women) from wage labour. In India also, large numbers of women were involved in 'heavy work', but the Colonial Government had to pay lip service to patriarchal values within India, and may have feared a loss of face if Indian women in Trinidad and Tobago were not seen to be 'protected housewives'. In India, the secluded wife of the twice-born castes was the symbol of highest caste and class status but, among agricultural labourers from whom most of the migrants were derived, this was and is definitely not the case. Thus, in his report, Comins (1893: 36) makes the general statement that 'Women are treated kindly both by their masters and their husbands, and do no work during the last months of pregnancy or after confinement. At other times, if they prefer to stay at home, no compulsion is used to make them work.' But he adds that 'of course during crop time it is expected that they will do all they can'.

As Comins (1893: 3) visited various estates he noted in his Diary that at Aranguez Estate 'I noticed a large number of women who had not gone to work. It seems free women are not obliged to work as their husbands and children earn plenty of money they stay at home and look after the house.' At Esperanza and Phoenix Park, Comins (1893: 20) noted 'Most women do only light work and some none at all; many do day work and some of the women prefer tasks.' These statements have to be seen in light of the fact that figures for the 1891 census, appendixes to that very report, classified only 6.2 per cent of Indian women as being involved in household duties. By creating divisions between 'heavy work' and 'light work', the planters were able to devalue women's labour; and by using the housewife/breadwinner model, divide the working class along sex lines. In addition, through this device the planters, as we shall see in the following section, could exploit the labour of two full workers but pay the wage of only one and maybe one half.

Wages and Remuneration

The discussion of wages is very important as, again, it lays bare many of the underlying assumptions governing women's labour. From the very inception of the indentureship system women were paid lower wages than men. The first one-year contract proposed by Governor MacLeod in 1845 included the wages shown in Table 3.2.

It is interesting to note how different categories of male labour existed while only one existed for females. In 1870 and 1875, the fixed minimum wage of 25 cents was set for all male able-bodied labour. Comins noted in 1891, however, that all women, weakly men and children were permanently paid a wage less than 25 cents. In this same report, where he said that women often did the same work as men, he justified the difference in wages by stating that it was 'because it has been decided that they are unable to do a full task'. Comins (1893: 9) continues: 'with this arrangement they are perfectly satisfied, for they know very well that the money

Table 3.2. Wages of Indentured Labourers 1845

Category of Labour	Wages in Equivalents			
	Rupees	Dollars	Sterling	
Sirdar*	Rs.7 per month	$3.35	13s	11d
Headman	6 per month	2.90	12s	1d
Male Labourer	5 per month	2.40	10s	0d
Female Labourer	3 per month	1.45	6s	0d

* Sirdar (or Sardar) is a leader.

Source: I. Cumpston, *Indians Overseas in British Territories*, Oxford University Press, London, 1953, p.99.

that they get is full value for the work they do'. Unfortunately, reports on the many strikes and protests of this period over wages and conditions do not testify as to the extent to which women were also involved in these actions. Thus Comins (1893: 18) found that at Woodbrook Estate, the wages for women's jobs ranged from 10 cents per hundred holes for manuring (pen) to 25 cents for weeding and planting cane. Men's jobs carried wages ranging from 25 to 40 cents. At Brechin Castle Estate free Indian men earned 30 to 50 cents while women, girls and boys earned 10 to 15 cents for manuring and supplying for the week ending 6 June 1881. At Palmiste Estate male factory workers could earn between 50 and 70 cents a day, but all women got 25 cents, and boys over ten years 15 cents a day (Comins 1893: 36).

About fifteen years later similar wage differentials were still in existence. The 1910 Sanderson Report in its Appendix gave the return of immigrants' wages for the period 1 April 1907 to 31 March 1908. These returns were important as the average male wage per day was used as the criterion for judging conditions on an estate. This list, unlike most others, included female wages, although they were not relevant to the issue. The Trinidad and Tobago Immigration Department never compiled adequate returns of wages although advised to do so by the Sanderson Commission (Lal and McNeil 1915: 19). An examination of these returns showed that the average daily wage at Brechin Castle estate for the year April 1907 to March 1908 for days actually worked was 25.33 cents for men and 20.64 cents for women. More importantly, when the average wage, per day per annum is examined, it is found to be 13.56 cents for men and 5.31 cents for women, the latter far less than 12 cents, the minimum stipulated for men (Sanderson Report 1910: 139–40). The wages of women, therefore, could deteriorate to any level without causing alarm since, based on the family wage concept, it was always assumed that a woman was dependent on her husband. What this situation did, in fact, was to force women in spite of the fact that they were full-time wage-earners, to become economically dependent on men, thus strengthening the men's power within the household.

But this was not all. Comins (1893: 15) also found that on some estates there was the practice of 'carrying forward . . . an ever accumulating debt for rations supplied to pregnant women'. This resulted in these women earning no wages for months, or even years. Of course during such periods, although working on the estates, women were totally dependent on the men. But, de facto, by their insistence on having the cost of the rations supplied during pregnancy repaid, the planters made it clear that they were not willing to actually pay a family wage or bear any cost of the production or reproduction of labour power. 'The family wage' therefore was/is nothing more than a useful concept for getting the labour of two and paying for one.

In his recommendations to the British Indian and Colonial Governments, D.W.D. Comins suggested that the period of indentureship for women should be reduced from five to two years. This move, he advised, would result in attracting to the colonies women of a more respectable character, as men would be encouraged to bring their wives. In other words, this was an attempt to alter the position of women emigrants from that of emigrant workers to dependent 'housewives'. Based on this recommendation, in 1894 the period of indenture for women was reduced from five years to three years, but in 1907 when the Sanderson Commission visited, it was found that there had been no apparent 'improvement in the number or class of women' (Sanderson 1910: 249). In British Guiana where a similar situation emerged, attempts were made to discuss the possibility of the free emigration of women. To support this, the following statement quoted in Ramnarine (1980) was made by one witness to the Sanderson Commission (1910: 112).

> In any country you like, if a man were told his wife would be indentured and have to work every day and all that, he would think twice before going to that country; whereas if he were a good working man himself he would be quite pleased if his wife came to the country to cook for him and work if she liked. I have often found that when women were indentured, they worked much better than when they came free. They thought they were independent and could work.

This quotation summarizes quite clearly and succinctly the 'have your cake and eat it' attitude which the planters hoped to exploit in the male Indians. They recognized that while the men needed the income of the women, they wanted that income to be earned under conditions in which they also have control. That this was in Comins' (1893: 49) mind as well is apparent in the concluding statement of his recommendation when he suggests that

> A woman who has been two years under indenture will be able to earn good wages in the field and will have acquired habits of industry, and there is little fear that when work is plentiful and good wages are to be obtained that her husband will allow her to sit idle.

By 1913, when the last two Commissioners, Chimman Lal and James McNeil came to Trinidad, the guidelines of this Commission included the investigation of the length and other conditions of the indenture of Indian women and the allegation that the large number of suicides and the prevalence of immorality on the estates were direct results of the indenture system (Lal and McNeil 1915: Appendix A, 326 – items 13 and 14). By this time, women were officially indentured for three years. In practice though, they continued to work for the following two years during which time they were reported to be better workers than during the first three (DPRO 1915–18: CO 295/467.29112). In 1913 the Commissioners found that women earned about one-half to two-thirds the wages of male workers, which amounted to between 60 and 72 cents weekly. They also found that some men were withdrawing their wives from wage labour so that they could 'devote themselves solely to domestic work including the care of children'. In the early twentieth century this practice was becoming more frequent for a number of reasons. At this period the colonial state was supporting Indian men in their attempts to recreate the Indian family life in Trinidad and Tobago. The redefinition of women as wives and mothers facilitated (as already noted) the local reproduction of the labour force, the 'stability' of the local population and most importantly provided *free* labour for the peasant production of cane and food crops. Thus men withdrew 'their' women from wage labour, but not to look after the house, as was officially stated.

Withdrawn into the domestic economy, many women assumed technical responsibility for cane farming, market gardening, rice production and the husbandry of animals such as chickens, sheep, goats and maybe one cow. However, they received no wages and were officially defined as non-earning 'housewives'. In this way, many women continued to work on the estates, combining this with peasant, market and subsistence production which also reduced the costs of the reproduction of labour power. The importance of this trend was recognized from its inception. Rev. John Morton, testifying in 1897 before the West Indian Royal Commission (WIRC 1898: 278) remarked that the depressed wages of that period were sufficient for an industrious man with a wife and children 'whom he can get to work on his own land or mind a cow'. Put so crudely, the economic advantage of having a wife and children was clear. It was apparently cheaper at that particular time for some men to do this than to allow their women to earn low wages. But what of the close to 50 per cent of Indian men who had no wives and hence no children? Their obvious disadvantage added a new dimension to the 'scramble for women' which was taking place at this time.

The importance of this development was again highlighted in the same WIRC (1898: 283) report in the complaint of H.A. Alcazar, then Mayor of Port of Spain, that peasant proprietors employed their own wives and children, and not outside labour which could benefit greatly. This domestic arrangement, however, was not possible for the majority of Indians, therefore many women continued to work on the sugar estates. In spite of the women's low earnings, Lal and McNeil (1915: 20–1) found that 'the best women workers earn almost as much as the average man'. Bearing in mind the low wages which they received, this must have meant that the women were doing more tasks than the men, yet in the Commissioners' recommendations no mention was made of women's wages. Instead, an average wage of 1 dollar and 30 cents weekly was recommended for able-bodied men (Lal and McNeil 1915: 21). Throughout the period of indentureship, therefore, in spite of the fact that over 90 per cent of all Indian women were engaged in agricultural labour and working under exacting conditions alongside men, they were paid lower wages; and like slave women before them, were excluded from the prestigious highly paid jobs. This created a situation in which these women, although wage labourers, were often forced to become economically dependent on men. Being in short supply, they sometimes had some choice in this matter, but in the last two decades of the century the combination of depressed economic conditions and consequent changes in the relations of production resulted in increased curtailment of the autonomy which they had previously enjoyed. The following section will discuss this in greater detail.

Social Organization and Indian Women's Struggle for Autonomy

The organization of life on the plantation was such that, on the one hand traditional practices from India had to be severely adapted or changed, while at the same time every effort was made to isolate the Indians from creole society. A number of important factors together contributed to the social organization among Indians during the indentureship period. These contributing factors can be identified as:

1. The organization and working of the plantation system
2. The breakdown of caste endogamy and of the patriarchal marriage and family system
3. The efforts to reconstruct it
4. The unequal sex ratio
5. The independent character of the immigrant women
6. Their struggle to retain their autonomy
7. The actions of the colonial state and the church in determining or supporting the emergence of certain forms of social organization.

During the initial stages, the plantocracy's primary concern was labour. This overshadowed all but the most minimal concerns for the reproduction of labour power. The workers were, for the most part, housed in barracks divided into two-room dwellings about 120 feet square. Each dwelling usually housed either two or three single men or one nuclear family of man, woman and their children (Comins 1893: 1; McNeil and Lal 1915: 4; Weller 1968: 58). One source suggests that during the early period houses were provided for families (Williams 1964: 105) but as the proportion of women was even lower than it was later on, this must have been quite negligible. Early regulations, according to Weller, prohibited cooking in the dwellings, but at the turn of the century, probably when more nuclear households were established, a cooking area was introduced to newer dwellings (Weller 1968: 59). With the shortage of women, men and estate management had to take much responsibility for the reproduction of male labour power. E.B. Underhill, writing in 1862, found that: 'the Coolies on the estates usually form themselves into messes, one man providing food, cooking it, and charging each man a fair proportion of the cost' (Underhill 1970: 66).

Based on the experience of the first indentureship period when, on arrival, large numbers of immigrants had died from malnutrition and overwork; after 1845 a system of rationing was introduced. According to Governor MacLeod's first one-year contract, these included: 45 lbs rice, 9 lbs peas, 3 gallon ghee or oil, 18 lbs salt, 4 lbs turmeric or tamarind, 1 lb onions or chillies per month and clothing – two blankets, two dhoosies (sic), one wooden bowl, one jacket and one cap per year (Cumpston 1953: 99). With the introduction of five-year contracts, these rations were given only for the first twelve months, until the immigrants were acclimatized. In return for these rations, 11 cents per day was deducted from their earnings. As a result of this, many workers became indebted to the estates when ill or pregnant and nursing or earning depressed wages. By 1869, the rations had been greatly reduced.

In 1870, according to Ordinance 13, the rationing period was extended to two years, workers paying 8 cents per day. Prior to this the one year's rations had often been delivered cooked. Now, the additional second year's rations had to be delivered uncooked (Weller 1968: 62). In spite of the administrative difficulties it posed, this practice had the effect of reducing the mortality rate and the rate of absenteeism, and served the planters' interest. The workers, however, resented its reduction of their cash income and protested against it. In 1894, the cost was reduced from 8 to 6 cents which was the real cost of the rations to the planters. Weller (1968: 73) refers to occasional instances of women being deprived of rations while their husbands were imprisoned. Such a situation might suggest that these women, if indentured, did not receive rations in their own right, or that rations were distributed only to male 'heads of households'. It is also possible that these women were not indentured and were therefore economically dependent on their husbands.

The breakdown of caste endogamy was another result of the plantation system. As already noted, although the majority of the immigrants were members of lower agricultural castes, small numbers of Brahmins, Kshatriya and other twice-born castes were also included. As would be expected, the members of the twice-born castes were those who had been forced through economic circumstances to seek a new life. For women in particular this was true, as widows of twice-born castes in India were often subject to a miserable existence in the homes of their in-laws, and were prohibited from re-marrying. Among the lower castes in India, however, which comprised the majority of the migrants, widow-remarriage was practised during this period of the nineteenth and early twentieth centuries (DPRO 1915–18: CO 571/5: 27270; Mies 1980b: 47–8).

The physical organization of the plantation, or before that, of the emigrants' depot in Calcutta, allowed little possibility for the observation of caste principles based on impurity and pollution. On estates, barrack housing, communal bathing and drinking facilities as well as communal labour broke down the remnants of the caste system, but it remained a sensitive issue (Brereton 1979: 185). By far the greatest challenge to the caste system came from the impossibility of using caste endogamy as a basis for marriage. With the shortage of women, caste could no longer be used as a criterion for choice. In addition, women and/or parents looked for men who could offer the best social and economic possibilities, and this did not always correlate with high caste. That it remained a sensitive issue was manifested by the fact that this was one of the underlying reasons for the non-registration of Hindu marriages. According to J.C. Jha (1975: 3), 'Those who had taken more than one wife or ignored the rules of caste endogamy would not like to expose themselves in public.' In addition, after serving their terms of indenture many Brahmins began to practise as pundits and assert their authority (Jha not.dated.I: 6). In disputes over inheritance, caste was again resurrected, often to deprive a woman and child of the property of a deceased man (Comins 1893: 31). Thus a submerged consciousness of caste continued to exist. In its place, some suggest, a kind of village endogamy developed, and other groupings in the society such as Muslims, Africans, Whites, might have assumed caste-like qualities based on racial, religious and colour criteria (Brereton 1979: 182–3; Reddock 1993).

To a large extent, the problem of the disproportionate sex-ratio could be said to have affected every other aspect of life. Throughout indentureship it continued to exist, and it was not until near the turn of the century that the ratio on incoming ships exceeded fifty women to one hundred men. Tables 3.3 and 3.4 illustrate the extent of the situation.

Table 3.3 shows that after a slight decline in the ratio of male to female immigrants in the last decade of the nineteenth century, there was a sizeable increase at the beginning of the twentieth century. To a large extent this was due to the stringent controls established during the latter period in an attempt to attract the 'right kind of women'.

During this time, as noted earlier, many prospective female immigrants were rejected if they did not have the permission of husbands or fathers, and many

Table 3.3. Sex Ratios (Males per 1,000 Females) among Estimated Net Immigrants to Trinidad (1871–1911)

Period	East Indian	Other
1871–81	2,143	1,101
1881–91	2,117	1,246
1891–1901	1,748	1,147
1901–11	3,037	744

Source: J. Harewood, 1975. *The Population of Trinidad & Tobago*, CICRED Series, p.102.

Table 3.4. Proportion of Women of every 100 Men among Emigrants from Calcutta to Trinidad 1891–1917

Year	Proportion of Women (%)	Year	Proportion of Women per 100 men (%)
1891	48.81	1905	42.02
1892	52.09	1906	39.22
1893	42.08	1907	40.34
1894	45.50	1908	40.04
1895	57.16	1909	39.42
1896	50.34	1910	40.13
1897	51.95	1911	38.76
1898	44.44	1912	41.06
1899	41.36	1913	39.65
1900	60.73	1914	45.05
1901	49.08	1915	58.97
1902	41.35	1916	49.22
1903	40.76	1917	42.02
1904	37.57	1891–1917	44.72

Source: R. Shiels, 1969. Indentured Labour into Trinidad 1891–1916. Appendix B, B.Litt. thesis, University of Oxford.

'undesirables' were disallowed. But when the plantations came under pressure the number of immigrant women was to increase again toward the close of the indentureship period.

In addition to the disproportion in emigration (Table 3.4), Comins (1893: 23) also found it in births to the extent of 1,083 males for every 1,000 females. But girl children in Trinidad, unlike those in India (even up to the present time, except among tribals), assumed a special importance. Sarah Morton (1916: 342) identified one instance where a father-in-law sold a girl nine times for money and goods and never delivered her, while Charles Mitchell in evidence to the West India Royal Commission in 1897, stated that 'a person with two or three female children here has very valuable property, because the men want wives' (Brereton 1979: 182). Child marriage from the age of ten was rife mainly because parents wanted their money as soon as possible. Some writers suggest that bride-price replaced dowry among Indians in Trinidad (Tinker 1974: 203), but among lower class/caste and untouchable Indians in North India, bride-price was and is, in fact, the norm. Women of these castes were economically productive labourers and not secluded and non-producing housewives like the women of the upper castes. That bride-price assumed greater currency than it had enjoyed in India is, however, true.

The disproportionate sex-ratio, in conjunction with the factors discussed in this section, was largely responsible for the difficulties which occurred in the reconstruction of the Indian family. In the first place, in the mid-nineteenth century the planters were in no way interested in importing families. Cumpston (1953: 69) noted that in the 1840s the initial inclusion of quotas for women was merely 'window-dressing' for 'public consumption' as the preponderance of males had been a major objection of the Anti-Slavery Society. Family migration implied costs for the reproduction of a new generation of workers and for maintaining non-able-bodied women during pregnancy and confinement, as well as other non-producing family members. It was only in the closing decades of the century that there was a definite change in policy towards the establishment of families.

In this effort to reconstruct the Indian family, the interests of the colonial state and the church coincided with those of the Indian men. But what kind of family did they attempt to build? For most lower class/caste men, emigration had been a means through which they were attempting to improve their class and, if possible, their caste status. In this 'sanskritization' process (the process of vertical caste-mobility) a family's highest aspiration was towards the family organization of the upper castes, with the secluded, non-earning wife whose only function was the provision of sons. In the migrants' efforts to reconstruct the 'Indian' joint family, however, church and state support was far less forthcoming. The western nuclear family with its dependent non-earning housewife was to be the model for modernization and westernization, and the basic economic unit of the society. Thus, an ideal of 'patriarchal' Indian family was supported insofar as it resembled its western counterpart. The joint family, as it developed in the Caribbean, was much more a collection of nuclear units than 'a group consisting of coparceners, the male co-owners and co-heirs of undivided family property handed down by ancestors' (Mies 1980b: 111). By the 1870s, therefore, having a wife had become a social and economic necessity, and controlling and keeping the one you had was an issue of life and death. In the words of Brereton (1979: 182), 'the possession of a wife was an important symbol of status and masculinity on the plantation, a crucial element in the husband's self esteem which he could ill afford to lose'. But for large numbers of women this resubjugation was not what they had come for. Evidence so far available suggests that their attempts to make a new life for themselves did not include their being re-encapsulated in the patriarchal Indian family. Unfortunately, the views and opinions of Indian women on this question are unrecorded. History can only judge from their actions which were the cause of much debate and discussion among the colonial and emigration authorities (DPRO 1915–18: CO 571/5–6).

Sarah Morton (1916: 324) for example, wife of Presbyterian missionary, John Morton, in a tone of extreme disapproval, recorded her experiences with Indian women.

The loose actions and prevailing practices in respect of marriage here are quite shocking to the newcomer. I said to an East Indian woman whom I knew to be the widow of a Brahmin, 'You have no relations in Trinidad, I believe?'

'No Madame,' she replied, 'only myself and two children; when the last immigrant ship came I took a "papa". I will keep him as long as he treats me well. If he does not treat me well I shall send him off at once; that's the right way, is it not.'

To which Sarah Morton commented:

This will be to some a new view on women's right . . . A woman who had left her husband because he had taken another wife, said to me in the calmest possible way, 'You know, it would not be pleasant for the two of us in one house.' 'And where are you now?' Unhesitatingly she mentioned the name of her newly-adopted husband. 'And where is your boy?' (Quite cheerfully) 'With his father' (ibid).

Other evidence of Indian women's attempts to maintain some degree of autonomy can be found in the pages of the *Port of Spain Gazette* (POSG: 1903) at the turn of the century. There, in the column 'The Police Courts,' reports were given of court proceedings for, among other things, 'wife harbouring,' declared a crime under the 1891 Immigration Ordinance. In one such, the case of Emanally versus Suambar, Piria the wife of the complainant (who of course was in principle defined as having been passively enticed away) very skilfully

denied living with him (Suambar) and explained her presence in his house by stating that she had been on many occasions lately subjected to the most brutal maltreatment at the hands of her husband and had on this account left his house and had gone to that of the defendant who had a sister who was her friend and country-woman. At this stage the matter was adjourned for a week to enable the defendant to cite his witnesses.

The advantages of this situation were that women had some choice in the establishment of relationships, and could, of their own accord, leave one husband for another or have relationships with more than one. This kind of behaviour on one hand was seen as downright immoral, and even some present-day historians have accepted unquestionably these colonial and religious value judgements (Ramnarine 1980: 6; Weller 1968: 3). On the other hand, this conscious, planned action on the part of the women was viewed as impassivity. The women were assumed to be helpless victims who were seduced or enticed away from one man to another, and to be incapable of independent, intelligent decision-making. Indian men, however, saw this independence of Indian women as a source of shame, and their inability to secure the loyalty of one woman on whom they could exercise power and authority in this colonial situation only added to their frustration. After

1916, the complaints of Indian men were voiced through the writings of the illustrious Mohammed Orfy in his numerous letters, written on behalf of destitute Indian men, to the Secretary of State for the Colonies, the Indian government and other authorities. They included passages like the following

> Another most disgraceful concern, which is most prevalent, and a perforating plague, is the high percentage of immoral lives led by the female section of our community. They are enticed, seduced and frightened into becoming concubines, and paramours to satisfy the greed and lust of the male section of quite a different race to theirs . . . They have absolutely no knowledge whatever of the value of being in virginhood and become most shameless and a perfect menace to the Indian gentry. (Mohammed Orfy, on behalf of destitute Indian men of Trinidad, see DPRO 1915–18: CO 571/4 W.I. 22518 of 1916)

The men 'of other race' referred to here were the white plantation officials with whom Indian women sometimes voluntarily, sometimes forcibly had relationships (Weller 1968: 67). According to most sources, sexual relationships between Africans and Indians were virtually non-existent. Up to 1870 no known cases of cohabitation existed possibly because of the Hindu association of dark colour with low caste and untouchability (Brereton 1979: 183). Yet Mohammed Orfy must have felt that the issue of mixed-race relationships was a strong point on which to draw public opinion to his side, or if not, the situation must have changed drastically by 29 October 1917 when he wrote

> Immorality is in the throes of proving detrimental to the integrity of the Indian gentry. Three-fourths of the females are more or less harboured publicly or privately as concubines and paramours of the races of other callings, such as Europeans, Africans, Americans and Chinese in goodly numbers are enticing the females of India, who are more or less subtle to lustful traps augured through some fear of punishment being meted out if not readily submissive as requested. (see DPRO 1915–18: CO 571/5, 60843)

In their efforts to subjugate Indian women and reconstruct the patriarchal Indian family, Indian men resorted to the weapon of violence. In this specific instance, the cutlass (machete) the main work tool of the sugar plantation was used. Murder and violence against women was common to all the areas of high Indian migration; Trinidad was not the highest on the list. Between 1859 and 1863, twenty-seven murders of women (wives or mistresses of the murderers) took place (Wood 1968: 154) and between 1872 and 1900, eighty-seven took place (Brereton 1979: 182). After 1900, the number appears to have decreased somewhat, so that between 1903 and 1912 only twenty-six were recorded (DPRO 1915–18: CO 571/4455 Enc. No.7). Weller (1968: 66–7) noted, however, that a murder was recorded only when a conviction was obtained, and usually insufficient evidence was available to convict the man. In British Guiana between 1885 and 1890, forty murders of women were committed, of which thirty-three were by husbands or reputed to be husbands (Ramnarine 1980: 20). In spite of this violence, it is clear that the women did not give up their autonomy without a fight. The numerous court cases resulting from breaches of the legislation against enticing away or harbouring the wife of an immigrant, and the continual letter-writing of Mohammed Orfy also bear testimony to this. This struggle and the 'loss of manhood' which was often alleged to be a consequence of it also resulted in self-inflicted violence, resulting in high suicide rates in all immigration territories (Kondapi 1951: 27).

Women's Unpaid Labour and the Transition to Peasant Production

In spite of the experience of slavery where women predominated in field labour on the plantations, during the early years of indentureship few women were recruited. As we have seen in this study, under the extreme conditions of wage labour during the early period, the planters and the male labourers themselves took responsibility for the daily reproduction of labour power. For the former, this meant the importation of new labour and the provision of minimum food and housing requirements. For the latter, it meant responsibility for their own domestic arrangements such as cooking, cleaning, and so forth. By the last two decades of the century, however, a change of policy could be discerned. This was the result of two main forces: 1) the economic crisis of the 1880s and 1890s caused by the increase in competition in the international sugar market; and 2) the increase in male violence against Indian women and the demands of male Indians for greater access to and control over their women.

The main solution implemented dealt with both of these problems. It was the encouragement of small-scale peasant and subsistence production of sugar-cane and food crops based on the labour of the family unit. Initially, hired labour was not to be used, and this change affected only one section of the Indian population. This new organization of production relations, achieved the following.

1. It reduced, for the planter, the costs of labour power reproduction.
2. It diversified the sources of cane for the factories, thus increasing the extent of labour control within and outside the estates.
3. It provided Indian men with the economic base for the reconstruction of the Indian family.
4. It increased the dependence of individual women upon individual men, although women continued to labour, by facilitating the transfer of women from public wage labour to more privatized peasant and subsistence production.

Woman Worker on the Caroni Sugar Estate, Trinidad. Trinidad and Tobago, circa 1986

Woman Worker on the Caroni Sugar Estate, Trinidad. Trinidad and Tobago, circa 1986

Women's labour was now seen as part of the domestic economy and an extension of their activities as 'housewives'. In a context in which public wage labour of women was often linked with low morality, this may have represented an improvement in status for men as well as some women. We do know, however, that acceptance of this new arrangement was not unanimous, as many women fought, in the face of much violence, to maintain their autonomy.

General Conclusions

To conclude this chapter, it can be said that for Indian women as for their African predecessors, the entrance into Trinidad and Tobago society as labourers, in spite of the difficult circumstances, presented the possibility of a break in the continuity of traditional patterns at home or male dominance over their lives. So long as it was profitable, plantation owners assumed much of the responsibility for the daily reproduction of labour-power and the reproduction of the labour force.

With changes in the economic situation, largely brought about by the international economy, the plantocracy adopted new strategies for managing women's labour and the production and reproduction of labour-power. In both cases these were a combination of production for the market by wage-earning workers and independent subsistence and market production carried out by the family unit. These strategies often reflected a coincidence of interest between oppressed men and the plantocracy or the colonial state, and were extended only to sections of the labouring population. However, they encouraged and promoted selective aspects of the European conjugal family and the colonial domestic ideology. As described in this study, these had to be adapted for women who would continue to be economically active. By and large the imposition of this strategy was not accepted by Indian women without a struggle; and because of their defiance, a much shorter (and less controlled) period was involved than in the case of slavery. However, more complete male dominance could be imposed through the reconstruction of a patriarchal Indian family, especially since this was supported by the colonial state and church.

4

Migration, Labour and Plantation Women in Fiji: A Historical Perspective

Shaista Shameem

Introduction

Fiji became a British colony in 1874 when, facing financial pressure and white settler lawlessness, the Fijian chiefs ceded the country, then under the kingship of King Cakobau, to the British queen. The first governor of the islands, Arthur Gordon, was ordered to make the colony self-supporting. He set up a system of native administration which acted as a form of social control. In addition, Fijians were forced, through tax obligations, to produce various types of commodities for the colonial state. The early 1880s witnessed a massive infusion of Australian capital and an increasing growth of sugar production. Since the state policies regarding native administration and a devastating measles epidemic had slowed down the flow of Fijian labour, the capitalists urged the government to find other sources of labour. Governor Gordon's solution to labour problems was to import indentured labourers from India. He drew upon his previous experiences in Trinidad and Mauritius where Indian indentured labourers had been working on the sugar plantations for several decades. Between 1879 and 1916, 60,965 people from India travelled to Fiji. (For further information see Sutherland 1984.) Approximately 28 per cent of the labour force were females. The biggest employer of indentured labour was the Colonial Sugar Refining (C.S.R.) Company of Australia.

From 1880 to 1916 sugar-cane was grown in large plantations which the C.S.R. either leased or bought outright from the native landowners through the colonial government. The relationship between the government and the company remained close. Both land and labour were made available in sufficient quantities to facilitate efficient production of sugar for export. There was no evidence of a partnership between the two. On the contrary, the company exercised tremendous power over the government. Whenever it looked as though the governor or some other colonial official might stand in the way of the company, C.S.R. would threaten to withdraw from its Fiji operations. Since the company provided most of the revenue in the colony, this was taken seriously (Sutherland 1984: 94). The 'coolies' were housed

49

in barracks or 'lines' on the plantation itself, a situation closely resembling slavery. The organization in the plantation and adjacent mills was strictly hierarchical. At the top was the C.S.R. manager, then came the overseers who were usually white Australians. At the bottom level were the sirdars. These were Indian men who were put in charge of small gangs of workers who were also divided into groups. When a new batch of workers arrived on a plantation they were sorted out into two main gangs. On the one side were the men who looked capable of doing heavy work, on the other were the weaker looking men and women. The stronger workers dug cane holes or laid railway lines, the weaker ones and the women were set to work weeding the plantations or helping to cut *tiri* (mangrove).

Generally, women had their own gangs and were kept separate from the men. Many women and some of the weaker men were also put to work in the flower and vegetable gardens and kitchens of the managers or overseers. Thus, the workers were divided hierarchically, not only in terms of their physical stature and work capability, but also along sex lines. The wage differentials also reflected the hierarchical nature of plantation organization. For example, mill work was considered to be skilled work for men, and paid more than field labour. Furthermore, women fieldworkers had a lower rate of pay than men. After 1890 when C.S.R. began to lease estates to its managers, the division of labour was maintained along the same lines. The other type of farmer, the Indian tenant farmer, employed additional male labourers when they were needed. The female members of the family worked in the vegetable gardens, but no longer in the cane field.

The first accounts of the role of women in the indentured labour force in Fiji were written by missionaries and philanthropists who wished to show that the colony was rapidly developing under the guidance of the British state and the Australian Colonial Sugar Refining Company. These Europeans also believed that the average Indian indentured worker who had been brought to work in the sugar plantations was benefiting from the system, and was able to accumulate money and possessions within five years. What did worry these writers, however, was the issue of plantation immorality.

Missionaries J.W. Burton and C.F. Andrews (who was Gandhi's close companion) were not so much concerned with examining the real poverty that existed among the indentured community, as with moral standards on the plantations. Both Andrews and Burton condemned the shortage of females in the labour force, and implied that those women who did come to Fiji were prostitutes. Andrews' (1918: 36–7) report, in particular, was widely circulated in India, Australia and Fiji, and caused an uproar in liberal circles. His report condemned the system of indenture on the grounds that it degraded and humiliated women. He said that the condition of marriage, so central to the Indian way of life, was impossible in Fiji where women were like 'rudderless' vessels with broken masts, drifting on the rocks. Indian women passed from one man to another, with no sense of shame.

It was the unmarried condition of Indian indentured women, and not their class position in Fiji as an exploited and subservient labour force, that horrified Andrews. His recommendation was that married women with families should be encouraged to come to Fiji, and that women of 'loose' character should be refused admission. Andrews also called for a new marriage ordinance and a ban on divorce, so that Indian married life would remain 'chaste and pure' as in India. The relationship between the Christian church in Fiji and both the company and the colonial state can best be described as contradictory. As a result of their evangelical work on the plantations, the Methodist Missionary Society of Australasia (MMSA) had the most contact with Indian labourers. The Mission ran the first orphanage for the children of labourers who had been murdered or killed on the plantations. It also set up schools and training institutions. On matters of morality, missionaries saw themselves as the experts, and frequently irritated the plantation managers with their criticisms of plantation immorality. At the same time, however, the mission schools for children of plantation workers allowed the C.S.R. to avoid setting up plantation schools. Instead, the company gave a yearly grant to the Mission for schools and hospitals.

The MMSA's relationship with the colonial state was also influential. The Society's involvement in education, health and political matters regarding Indians, enabled the state to present a respectable face to the rest of the world, as far as the condition of Fiji Indians was concerned. This was particularly so from 1914 onwards, when international attention was being increasingly focused on the indenture labour system. However, although the Methodist Mission in Fiji was frequently critical of both the company and the colonial state, it never challenged either colonialism or capitalism. For instance, Andrews' accusation that the immorality on the plantations was caused by the indentured labour system was not supported by all the Society's members. Rev. W.R. Steadman, writing to Rev. L.M. Thompson in 1881, said that the picture of Indian life in Fiji given by C.F. Andrews was not quite a true one. He said that the immorality of the Indians was caused by the disproportion of the sexes, and that immorality existed in India too (see MMSA 1914–23: 1, Fiji District Misc. Section M/33).

The evidence shows that the Methodist Mission in Fiji tended to defend government and company policies rather than challenge them. The Mission in Fiji depended on grants from both the state and the company for the running of their schools and hospitals. With the benefit of hindsight it is possible now to criticize Andrews and other missionaries like him, who sought to deal with exploitation from a moral standpoint; but they formulated opinions which even the more contemporary writers have recently echoed. Local writers in the 1960s and 1970s claimed that it was the severe shortage of women on the plantations that caused the moral downfall of Indian women (Ali 1980; Naidu 1980). They stated that women on plantations took advantage of their scarcity value and bartered their bodies to whoever paid the highest price. According to these writers it was the dreadful shortage of females which caused Indian indentured women to deny marriage and lifelong faithfulness to their husbands. Had the women been permitted to marry and have children, all the social problems on plantations would have been solved. An attempt was made by a local scholar (Lal 1983 and 1985a, b) to justify the moral condition of the women by collecting data on the caste and social origins of the female labourers. However, he reached the conclusion that one could not tell, from the data, whether the women who came to Fiji were from the prostitute class in India, as Andrews had maintained in his reports. According to the evidence collected by Lal, the women constituted as diverse and differentiated a group as the men.

The historical interpretations referred to above do not tell us exactly what the position of the indentured women was, relative to men. Why were Indian women brought out to labour in the field? Were they, like men, mere agricultural implements, or were they crucial to the reproduction of a future labour force? What particular contradictions did they face with men workers? Of especial concern is the assumption of the marginality of women in capitalist production, which appears as a central strand in much of the writing on the indenture system. There are few references to the actual position of women in paid employment. The fact that women were contracted to work on the plantations as field labourers for 9 pence a day, and that many continued (though in decreasing numbers) to work as hired labourers even after their indentures expired, has so far been marginal to the analysis of political and economic aspects of indenture. The androcentric biases in these interpretations have so far resulted, not only in the wrong questions being asked, but also in an assumption of marginality of women in production.

This chapter attempts to deal with the issue of marginality of women in capitalist production. It is divided into two main parts. In the first part, two key theoretical issues are examined with regard to the labour participation of women: (i) The implications of viewing women's labour participation as a product of both class and gender struggles, and (ii) how women's status and authority have been historically determined in the context of class and gender relations on Fiji's sugar plantations. Included in the second part is a brief reference to the resistance of women, both to capitalist exploitation and to male dominance, evidence of the fact that women cannot be seen as passive victims, but as living, struggling beings, capable of making their own history. This view challenges the commonly held view that Indian women were outside, or marginal to the capitalist economy, both during and after indenture.

In the debates on sex and class that have appeared in various academic journals over the past ten years, the central focus appears to be either an attack on, or a confirmation of Engels' (1978) prediction that women's emancipation will be assured once all women enter the sphere of public production. What has also been widely discussed is Engels' further point that the relationship between the family and the mode of production is crucial to an understanding of women's oppression. In their quest for the material roots of women's oppression, Marxist and socialist feminists have stressed the importance of recognizing that the gender system is as deeply ingrained in social and economic formations and the political institutions to which they give rise, as class relations. Furthermore, it is this gender system that has historically undervalued women's involvement, not only in social and economic development and in class struggle, but also in resistance to male oppression (Fox-Genovese 1982).

The feminist scholars who have explored the analysis most fully have been those engaged in the domestic labour debates. Two main trends can be identified in this debate. The first aims to show that the housewife's labour, under capitalism, makes an economic contribution to the capitalist system by reproducing and maintaining the male workers (Benston 1969; Dalla Costa 1973; Gardiner 1975). A second, which acknowledges women's contribution to the reproduction of capitalism, says, however, that women's labour in the home benefits men first. Men receive directly, the product of women's labour (Hartman 1981). Both Engels' point (that all women must enter public industry in order to be free), and the domestic labour debates, are relevant to the study of Indian women workers in Fiji, and raise the following questions.

First, all Indian women who came to Fiji to work under indenture were thrust into public production. But was women's entry into public production a sufficient condition for their emancipation on plantations?

Secondly, both men and women were labourers and worked under the same general conditions. In addition, because of the limited number of women, men could not expect that household chores would be the burden of women alone. Did this mean that relations between them were organized on an equal basis?

Indian Women as Plantation and Farm Workers in Fiji

During most of the indenture period (1879 to 1920), two systems of employment existed side by side. One was the plantation system and the other, the small farm system. In both cases the labour of indentured labourers, both male and female, was exploited by the Australian capitalists. The capitalists' desire to make a profit from sugar production allowed for the transportation and indenturing of both women and men for a period of five years. After their indentures expired, most Indians settled on leased farms and planted sugar-cane for the Colonial Sugar Refining Company's mills, or became waged labourers who depended on the mills and farms for employment.

Over the years, women were gradually displaced from the wage relation. After their five-year indenture period, women labourers re-engaged only in very small numbers. The large majority became involved in the domestic sphere where they produced for domestic and subsistence needs. The women's displacement from the wage relation, however, did not mean that there was a radical alteration in their links with the capitalist mode of production. They became even more subservient to capitalist relations by both providing wage labour when they were needed (thus forming a reserve army) and by contributing to domestic and subsistence production. Women's changed status – from indentured labourers to domestic labourers – transformed the nature of their relations with Indian men.

In the case of Fiji, both men and women were recruited for agricultural work. By the time Fiji was ready to import labour from India, the legal proportion of female labourers had been set at 28 per cent of the total number of adults. The C.S.R. company in Fiji continued to agitate for a reduction of this proportion on the grounds that only a small amount of work could be extracted from women (ARAGI 1896). The European planters, who had originally grown cane for the Colonial Sugar Refining Company and other millers, were gradually replaced by the free, Indian small farmer. The Indian family unit grew sugar-cane for the mills. In 1895, about 300 acres of land were under cultivation by Indian farmers (ibid). By 1917, the acreage under cane cultivation by free Indian contractors, or farmers in company settlements had increased to 1220. Since there were two systems of production and exploitation of labour operating at the time, namely, indentured and free, the position of women has to be analysed separately, as their relationship to production was different in each case.

Indentured Women

In the early years, when indenture was first beginning to take root as a method of labour organization in places such as Mauritius and the West Indies, few women were recruited as labourers because the planters and millers required only able-bodied male labourers. Women were requested only when the cooking and sexual needs of the male labourers had to be met. In 1836, when the earliest arrangements were being made for transportation of coolies to some of the colonies, John Gladstone, the proprietor of some estates in Demerara (formerly known as British Guiana and now Guyana), stated that indentured women should form a part of his workforce only if they were prepared to undertake field work, in which case they could form 40 to 50 per cent of the total; but if not, then one female to nine or ten males for 'cooking and washing is enough'(Tinker 1974: 63). If women could not provide their labour in the field with men (and many white planters believed that women could not) then only a few were needed for the reproductive tasks[1] traditionally carried out by women.

However, planters and millers quickly discovered that the women had some uses. Not only could they be put to work as agricultural labourers, but they could also be relied upon to provide domestic labour for the male labourers. In Fiji, the manager of the C.S.R's plantation Vucimaca was censured by the Agent General of Immigration for forcing a woman called Baggia to live with a man and 'cook his rice' (CSO 1909–16: 86/1194). The woman committed suicide because she did not want to live with the man, and the Colonial Secretary's Office sent out a memo to the General Manager of the Company, warning that managers and overseers should not meddle with the matrimonial arrangements of labourers. The female labourers were treated in the same general way as the male labourers, except in two important areas – work and wages. A strict sexual division of labour operated on the plantations, whereby women worked only in the fields and men had the option of mill work.

Since women never worked in the mill, their wages were based on the task system. The development of the plantation hierarchy based on levels of skill allowed for the involvement of women in unskilled field labour only. Mill work, with its better working conditions and higher pay rates, was reserved for men. In the early years of the system as it operated in Fiji, if a man was found to be unable to do a male's task, he was put on women's wages of 9 pence per day, and was expected to complete only three-quarters of the task. Later, the concept of 'limited tasks' was introduced and 'unfit' men and women were expected to complete only half or three-quarters of a task.

Work Conditions

The 1878 Indian Immigration Ordinance stated that the labourers would be employed in task work or time work for all days of the week except Sunday. At time work, immigrants were to work for nine hours: a task was defined as the amount of work which an ordinary able-bodied adult male could do in six hours of steady work. A woman's work was three-quarters of a man's task. In a week, it was expected that a labourer would be able to do five and a half tasks. Although the ordinance allowed for both task and time work, it was not the labourer who could choose which type of work he or she wanted. Only the employer could make that choice. Generally, field work was based on the task or piece rate system, and millwork was based on time rate. The stipulated rate for time work was not less than 1 shilling (Sterling) per day for a man aged fifteen or more, and 9 pence (Sterling)[2] per day for a woman not less than fifteen years of age. Those workers who were under sixteen (defined as non-adults) were paid in proportion to the amount of work that they did. Task work was paid at a rate of not less than 1 shilling for each task that the male worker completed. Women were entitled to 9 pence for finishing three quarters of a man's task. The wage rates remained the same for the duration of the indenture period (1879 to 1920).

In terms of field work, the following average tasks were expected to be completed by labourers: digging 150 sugar-cane holes (24'x18'x12'); weeding and trashing 10 to 15 chains, 6 feet wide; draining 250 to 270 cubic feet; cutting 3 tons per day; and shovelploughing 7 to 10 chains. The Stipendiary Magistrate in Rewa, Walter Carew, told the Colonial Secretary that the tasks set by the employers were too heavy (CSO 1909–16: 86/1268). However, there was no attempt on the part of the government to compel the capitalists to reduce the tasks. The intensity of work required from the labourers is demonstrated by the fact that during most of the years of indenture, the government expected that 25 per cent of the potential utility of those introduced could be lost in five years of indentured service. This loss was either through death or repatriation to India through incapacity. Within the year of arrival, 4.10 per cent of the adults introduced died. The figure for children was much higher. Within one year of arrival, the percentage of children lost through death alone was 19.21 per cent. This large loss was remarked upon by the Agent General of Immigration in his Report (ARAGI 1894: 470). The loss of children was seen by him to be detrimental in terms of the 'material' future of the colony, and he called the attention of the employers to this fact. The high rate of loss for adults was seen to be particularly serious since most of the adult labourers were suited to agricultural work, having been recruited from the agricultural classes, and were compelled, under the regulations, to undergo two medical examinations for fitness – one in India and the other in Fiji. In 1905, the percentage of adult females lost, in proportion to their population was 6.5 per cent, compared to 3.9 per cent lost in the male population (ARAGI 1905).

The fact that the percentage of days lost through legitimate causes (for example, sickness) was greater for women than for men, may have been the greatest factor influencing the attempts of the employers to reduce the proportion of females (see Table 4. 1).

The reason given by the inspectors of immigration for the high percentage of days lost by women in comparison to men was that the attendance of women was affected by pregnancy and nursing. It is clear, however, that given the extremely low wages of women, there was inevitable breakdown in health and well-being due to inadequate nutrition. There is no doubt that few adults ever made the

Table 4.1   Percentage of Days Lost by Males and Females through Sickness, Pregnancy and other Lawful Causes

	1885	1891	1895	1900	1905	1910	1915	1917
Males	19.24	12.65	8.37	5.42	2.78	5.91	2.31	2.72
Females	34.39	24.65	23.73	10.69	9.25	8.10	17.18	17.03

Source: ARAGI (Annual Reports Agent General of Immigration) 1886 to 1921

minimum wages stipulated under the regulations. According to Narsey (1979: 84) the greatest increments to profit for the C.S.R. arose from the savings in labour costs. The wage differentials between sugar plantations in Australia and Fiji were huge.

An independent Commission of Enquiry in Australia, looking into wages of Australian mill and field labourers, stated that the wages of about 10 pence per week were inadequate (Narsey 1979: 95). In 1912, Indian male labourers were receiving approximately one-fifth of the wages paid to white sugar workers in Australia. It was assumed that it cost Indian workers less to reproduce themselves than white workers. The report of a Commission set up to look into the Cost of Living in the Colony in 1920 revealed that the cost of food items purchased by the Indian labourers had more than doubled between 1915 and 1920 (FLCP 1920). Yet the rate of wages had remained the same. It also noted that the cost of clothing had increased even more dramatically in the same period. Most of the calculations of contemporary scholars, with regard to the wages of the Indian labourers compared to the cost of living, are based on the wages of the male labourers. The andro-centrism inherent in this analysis is highlighted when one examines the wages of female labourers in relation to male wages. Although the minimum wage for women per task or per working day was supposed to be 9 pence, it is clear that more than half the females never made this wage during the entire indenture period. In 1883, for example, in three districts, Rewa, Navua and Tavenui, only one-quarter of the men made their minimum wage of 1 shilling per day, whereas only 8 per cent of the women made 9 pence. The rest of the women earned less than that, or slightly over 5 pence per day. In the following year, 56 per cent of the women earned less than 5 pence per day (ARAGI 1884).

The fact that women's average wages did not even reach the stipulated 9 pence a day until 1917, nearly forty years after the indentureship system began in Fiji, and just before it ended, shows that women labourers suffered a greater level of poverty than their male counterparts. It also shows that the capitalists expected the women to reproduce themselves at a lower cost than men. No wonder the women lost a greater number of days through sickness and ill-health. This may also be the reason why so many children died in the first year of birth. The high death rate of plantation children (83.75 per thousand as late as 1913) was blamed on the 'care-lessness and indifference' of the mothers (Gillion 1973: 107). It was also suspected that many women killed their children, although this could not be proved. The main causes of death of children were diarrhoea, malnutrition, debility and con-genital syphilis.

Although many births and deaths, particularly of infants, were not recorded by time-expired Indians living off the plantations, it was obvious that the mortality among the 'free' population was much lower. It appears that those living on the sugar estates, whether indentured or under re-engagements, suffered greater losses than those Indians in settlements. The women's low wages, and consequently their inability to care adequately for themselves or their children, forced them into marriage, or casual associations with men. In 1915, the bare minimum that the state functionaries and capitalists thought was necessary for Indian labourers to live on was rations that cost between 3 shillings and 3 shillings and 6 pence per week. In the same year, the average Indian female labourer earned just over 6½ pence per day. Assuming that the Indian female worked for the whole five and a half days in the week as required, and not taking into account the high absentee rate, it becomes obvious that the Indian woman could never make her cost of reproduction, which, as set by the colonial officials, was the barest minimum (Narsey 1979: 86). In addition, between 1901 and 1906, although the attendance at work was higher, the wages the women earned were lower than in previous years – implying overtasking.

The women's substandard living conditions should be considered in terms of the reputation the Indian women of Fiji gained as prostitutes and 'fallen' women. The perspective of four writers C.F. Andrews, Ahmed Ali, Vijay Naidu and Brij Lal, regarding this point is pertinent to the discussion. C.F. Andrews (1918) stated that the plantation conditions were detrimental to the Indian women's chance of marriage because there were so few of them, and they moved from one man to another whenever it suited them. He recommended that a better class of women be recruited, and also argued for family, rather than single migration, on the grounds that it was within the family and the community that the women's chastity, so valued in Indian society, was protected.

Ali's (1980) position regarding indentured women is that they took advantage of their scarcity value, as men had to compete for the relatively few available. This view is echoed by Naidu (1980) who provides oral testimonies of women that any female indentured labourer had relationships with two or three men simultaneously. It is implied that the evil conditions existing on the plantations caused women to abandon traditional roles and become promiscuous. Lal's (1985b: 65) view that women were unable to earn a proper wage and were forced into prostitution can be supported by the fact that many Indian men preyed on the women's vulnerability and made fortunes bartering their bodies in exchange for payment of labour fines or bail money (CSO 1909–16: 86/617). The women accepted this option knowing that they could not survive the harsh plantation conditions.

Here, the analysis of work and wages of indentured women in capitalist production is significant for two reasons: Firstly, implicit in the capitalist organization of the labour process lay a fundamental assumption that women's needs were secondary to men's needs in terms of nutrition, and therefore that women require less to reproduce themselves. Secondly, a hierarchical sexual division of labour existed on plantations which not only barred women from skilled and better paid jobs such as mill work or supervision of work gangs, but also nudged them into traditional work roles of wife, mother or concubine. In other words, although there was a possibility of women becoming emancipated and free as Engels suggested, the reality was that the sexual division of labour, as it existed on capitalist plantations, did not permit this to take place. Engels, in his analysis, did not take into account the necessary sexual division of labour within capitalism. He also overlooked the power of males, who even when themselves subservient, were capable of forcing women into an inferior position. The effect that this plantation labour process had on the relationship between men and women is therefore particularly significant and will be dealt with next.

Relations between Men and Women on the Plantations

During the period of indenture, slightly over one-third of the indentured women were reported to be part of families (Lal 1985a: 57). This means that the rest arrived as single emigrants. The fact that many of the family associations may have been rather hastily established in India is shown clearly in reports of women's refusal to continue cohabiting with men to whom they were supposed to be married. In 1894, for example, several indentured women from the ship *Hereford* abandoned their husbands, insisting that they wished to part company. In one such case, the abandoned husband had to be placed under restraint. Finally, at his request, he was not sent to the plantation to which his former 'wife' was allotted (ARAGI 1894).

There is no doubt that the services indentured women provided for the men they lived with were welcomed by the capitalists. Women not only cooked and cleaned for their male partners, they also had a cushioning effect on overworked men who could not take out their anger on the all-too-powerful company. Women received the brunt of men's disillusionment and despair. The value of women's unpaid labour was recognized by the capitalists because they frequently allowed the women to commute their indentures when their husbands' indentures expired, and they re-engaged for a further five years. In some cases, a married woman served only three years of her five-year term. In 1881, an Indian woman who followed her husband (a policeman) to Rewa, was required to have her cost of indenture (i.e. transportation and food costs) paid by her husband. It was felt that since her labour was not required by the police department, and therefore her cost could not be paid by them, it was her husband who must pay for it, 'he being no doubt the only person who is receiving the benefit of his wife's services' (CSO 1909–16: 85/1520).

Single Women with Children

Single women with children were considered to be a liability by the colonial officials and the capitalists. In 1888, an unusually high number of single women with families came on the *Hereford*. Thirty-four women came unaccompanied by husbands, bringing with them fifty-one young children. According to the 1888 Report of the Agent of Immigration, the prospect of women earning a livelihood under such conditions was 'faint in the extreme'. He thought it was unlikely that the men would form alliances with them, burdened as they were with so many children, whose maintenance fell entirely upon parents after twelve months in the colony. In the first year, the children were looked after by the company employing the parent or parents. At the end of twelve months the mothers unassisted by husbands found it impossible to put away money after supplying their own wants and those of their children, still too young to work. Under such circumstances these women were prone to resort to means of livelihood which rendered them independent of the employment secured to them by indenture, and moreover resulted in their becoming unfit for it. They were soon confirmed non-workers, frequent inmates of the hospital and gaol, a serious expense at first, and afterwards a source of much trouble and inconvenience to their managers (ARAGI 1888: 1–2).

The Agent General recommended that single women with children should be discouraged from emigrating to Fiji. Childcare was inadequate on the plantations and women were forced to stay in the barracks to look after their children, instead of going to work. Thus, an indentured woman was burdened not only with lower wages and heavy field work, she was also required to cater for the domestic needs of husband and/or children. For this, she was not paid. In addition to the double burden of plantation labour and household labour, she faced the brutality of both overseers and Indian males.

Plantation Brutality and Women's Resistance

The subject of the violence of indenture has been dealt with adequately by authors such as Vijay Naidu and Brij Lal. What has largely been left unexplored, however, is the notion that indentured women faced violence and brutality from both the capitalists and indentured men because of their resistance to exploitation and domination.

One significant feature of the indentureship system was the penalty clause in the Immigration Ordinance. It stipulated that indentured labourers could be prosecuted and, upon conviction by a magistrate, fined or imprisoned for a wide variety of labour offences including desertion, unlawful absence from work, failure to show ordinary diligence or to complete a task; and offences related to discipline, such as using insulting language, or disobedience. An extension of service was made by the court to cover the time lost through absence or non-performance of task, the day spent in court and the period of imprisonment (Gillion 1973: 118). In most years of indenture, women were convicted, and their services extended in fewer numbers than male labourers. But as exploitation increased, the proportion of women sent to gaol, fined or forced to work after their indentures had expired exceeded the male proportion. In 1896, 1897, 1898, 1899, 1900, 1901, 1902 and

1907 there were more charges laid against women than against men. In 1898, for example, 44.2 per cent of the men and 56 per cent of the women were prosecuted. According to the Annual Report for that year, more than two-thirds of the charges laid against women in the colony were instituted against two hundred and six women in Macuata district. This district, dominated by the C.S.R., was notorious for overtasking workers. Although the women there had a higher work attendance rate in 1898 compared with 1897, their wages dropped by about 1 penny per diem (ARAGI 1897 and 1998).

Women's resistance to capitalist exploitation took the form of withholding their labour, but at the more personal level they also carried out physical acts of violence upon male overseers or sirdars. Women would get together in gangs and beat up cruel and abusive plantation personnel, utilizing hoes, knives and other work tools as weapons. They inevitably went to gaol for acts such as these (ARAGI 1901 and 1902). Women's resistance against domination by Indian males was also common on plantations. Some men received an early taste of this when they disembarked from the labour vessels. The women they thought were their wives because they had cohabited with them in India and on board the ship, refused to continue living with them. This was contrary to all traditional expectations regarding the sanctity of marriage and the lifelong faithfulness of the wife.

Further evidence that many plantation women resisted the dominance of Indian males is provided by figures for the crime of murder. Between 1890 and 1921, sixty-eight indentured women were murdered, compared to twenty-eight indentured men (Naidu 1980: 61). It was widely believed by the officials that murders of women were committed due to sexual jealousy on the part of males who could not cope with the immorality of the women. In murder trials in the colony during this period, evidence was frequently given that the female victim had either left her husband for someone else, or had threatened to do so. It was often the case also, that a woman took lovers, and when discovered, she was brutally killed. The men were thus supposed to have saved their honour.

The concept of honour or *izzat* is always pointed out by writers discussing the crimes against women on plantations (Ali 1980 and Naidu 1980). The androcentric assumptions inherent in these attempts to justify violence against women by pointing to the loss of male honour caused by women's immorality are obvious. There is no doubt that the brutality women experienced was caused by men's refusal to let them control their sexual lives. This was the case, not only in relation to Indian men, but white men also forced their attentions upon women, and those who resisted found that they would be given harder tasks the following day.

An additional aspect of violent abuse of women by men is shown by the figures for sexual offences on plantations. Between 1885 and 1920, two hundred men were charged with committing sexual offences. The majority of the cases were for rape and carnal knowledge of girls under thirteen (Naidu 1980: 70). Since a large proportion of cases for sexual offences usually goes unreported, the real figures for this type of crime were probably much higher.

Finally, there were countless cases of assault and battery of women. These cases were so numerous and so much a part of the general brutality of plantation life that most of them did not even reach the courts. Women were therefore recipients of both European capitalist and Indian working-class male brutality. They resisted dominance and paid the price – in many cases with their lives.

'Free' Women

In the written history of indentureship in Fiji, there are frequent references to the changing nature of women's lives after their indentures expired, and as they settled on small farms with their husbands and families (Gillion 1973; Mayer 1961; Jayawardena, C. 1975). There are even underlying assumptions, in the more recent literature, that the ideal place for a woman is with her family. For example, Lal (1985a: 138), in a recent paper on suicides suggests that rather than blame women's immorality for male suicides on plantations, one should look for underlying causes, for example the collapse of integrative institutions of society such as family, kinship, marriage, caste and religion. Certainly there is little evidence in the records to suggest that the major motive for male suicide was women's promiscuity. However, in seeking to discredit this widely held male-biased view, Lal himself falls into the trap of androcentrism. He fails to see that the 'integrative institutions' themselves may be detrimental to the well-being of women. His interpretation also does not explain why the failure of these institutions should result in men and not women committing suicide. Surely women were not immune to the alienation caused by the collapse of culture during indenture.

The domestic labour debates discussed earlier in this chapter point to the importance of analysing the position of women in the family, because it was within the family that the women laboured for free. This point is of crucial significance as far as the conceptualization of Indian women in Fiji is concerned, for certain aspects of 'free' family life were as carefully orchestrated by the capitalists, as were relations between indentured men and women. Women were not given leases to farm, and they were not encouraged to reindenture after their indentures expired (Jayawardena, C. 1975: 10).

It must be remembered that the period of indenture for an individual was at least five years, and at most ten. After the indentures of Indians expired, most decided to settle near the mills in the cane growing areas. Many were able to lease land and plant cane themselves, while others worked as free wage workers (for about 1 shilling and 6 pence per diem) on the large estates and surrounding hills. During indenture, therefore, two different systems of production operated – one, the plantation system, the other, the small-farm production unit. Up to 1912, the small-farm system was seen mainly as a means of supplementing the supply of labour on the plantations, but after that year, it began, slowly, to replace the plantation system of production. By 1918 it had become obvious to planters that the indenture system would soon be abolished.

The restructuring of the production process by the Colonial Sugar Refining monopoly went through several stages, but eventually tenant farmers were leased about 8 to 12 acres of land each. This amount was supposed to be ideal, in terms of the utilization of the labour of a male farmer and his family. 'Indeed, the nuclear family was the tenant the Company preferred. When farms were originally allocated, only married men were, as a matter of policy, selected.' Furthermore, the company did not leave the running of these farms to the tenant farmers. Sutherland (1984: 106) points out that the major form of surplus appropriation in the small-farm sector involved control over both the price paid to the farmer for cane, and the methods of production. While the price paid for cane was insufficient to maintain white living standards, it was expected that Indian farmers would be 'able to extract a good livelihood from the cultivation of cane' (Sutherland 1984: 108). The company's control over production methods involved regulating cultivation, surveillance by company officials, control over harvesting and so on.

Sutherland (1984) and Narsey (1979) correctly point out that the expectation that the Indian tenant would accept a lower price for his cane than would a white planter, was based on racist assumptions. But there was also an assumption that the subsistence needs of the producers would be met through the cultivation of family vegetable and rice plots. Since it was the man who was responsible for cane cultivation and harvesting, it was inevitable that the woman would become the subsistence gardener. She could combine this with her household tasks. This sexual distinction had its roots in the division of labour brought about by capitalist restructuring of the production process.

It is at this point in history that women begin to fade out from official labour statistics. Instead of continuing to be categorized as agriculturalists or farm labourers, they are reported as being 'engaged in domestic duties'. This is clearly illustrated by comparing two Fiji Census Reports (see FCR 1911–21), 17.9 per cent of those women workers who are not indentured are categorized as agriculturalists. By 1921, only 1.8 per cent of women workers are in the 'agriculture employment' category. In the category of domestic workers, in 1911, 35.4 per cent of the women were labelled as domestic workers, whereas in 1921, this was the occupational category of 88.5 per cent of the women over ten years of age.

Statistics such as these are frequently used by development planners, and even academics, as evidence of the fact that women are marginal to production. Fiji's Indian population, after indenture, continued to be rural based; and in this setting it would be impossible for women to be 'marginal to production'. The women whom the Census Reports categorized as 'having no occupation' or as 'involved in domestic duties' were, for the most part, engaged in the double burden of production for home consumption and reproduction of the next generation of workers for the sugar industry of Fiji. The fact that the capitalists benefited from the small-farm system is shown by the amount of money the C.S.R. made in Fiji. Narsey (1979: 98) estimates that over the ten-year period between 1914 and 1923, the company made 'superprofits' of about £13 million.

The relations between men and women on the small farms could not help but be influenced by this new structure of public and private. Although many writers see this relation as being complementary and equal rather than antagonistic, I do not share their view. The division of labour between the sexes on Fiji's farms has never been equal. Men do not merely dominate the public sphere, they also claim seniority over women in the household. Neither are resources equally shared. Women may have access to household resources but the control rests with men. In fact, the institutions of marriage and the family, contrary to popular belief, have reinforced the subservience that women experienced during indenture. Women are doing the same amount of work, but this time for no pay.

Conclusion

There are several significant aspects of women's condition, under indenture and on small farms – such as religion, caste, culture and ideology. All are important, and work has yet to be done on them. The aim of this chapter, however, was to deal with the labour of women, following Engels' position that 'according to the materialistic conception, the determining factor in history is, in the final instance, the production and reproduction of immediate life – on the one hand the production of the means of existence, and on the other, the production of human beings themselves'. I am concerned here with this 'two-fold' character of production and reproduction of material life in a specific historical context. In the case of Indian women, it is true that the possibility of their emancipation came with their entry into social production. After all, as waged labourers they struggled against capital with their male counterparts. However, the fact that they were supposed to exist on less pay than men because they were women, undermined their chances of emancipation. They were forced to rely on men for their subsistence and for their children's needs. In addition, they faced male brutality (as predicted by Engels), which further reduced their chances of freedom. After indenture, although women's participation in production remained as important, its lack of recognition, as well as the nonpayment for agricultural tasks, ensured their further dependency on men for survival. This situation has remained largely unchanged for the majority of rural Indian women in Fiji today.

Notes

1. The term reproduction is used here to refer to the daily maintenance of the workers. The concept is also used to refer to human or biological reproduction, and to the allocation of agents to positions within the labour process over time. See for example discussion by Edholm, Harris and Young (1977).
2. In 1878 one shilling was equivalent to about 10 US cents.

5

Tamil Women on Sri Lankan Plantations: Labour Control and Patriarchy

Rachel Kurian

The Origin and Development of the Plantation Workforce in Sri Lanka

The development of the plantation economy in Sri Lanka was closely connected with the needs of industrial capitalism in Britain in the nineteenth century. This called for cheap raw materials and other inputs, such as food and beverages, to keep the cost of production in Britain as low as possible. Colonies were encouraged to promote products which would be important in this respect, and at the same time be sources of wealth in themselves. It was in this context that coffee, tea, rubber and coconut plantations were developed in nineteenth-century Ceylon; and came to provide the basis of capitalist expansion in the country. One of the most important problems facing the development of these plantations was how to acquire and control an adequate supply of labour to meet the production requirements. In the early days, this was not forthcoming from the Sinhalese population which inhabited the villages surrounding the plantations. This was largely due to the fact that they were tied to the land; and the early alienation of land for plantations did not interfere with the ownership of land or with feudal work systems prevailing in villages (Jayawardena, L. 1963). As a result, planters turned to alternative sources of labour for the plantations. And this they found in the Tamil districts of the Madras Presidency in South India. Labour from these regions had already been employed in the building of public works and in servicing officers in the island (SLNA 1840). In the initial stages, due to the proximity of India, work was generally of a temporary nature, and workers would migrate for a season and return to their homes when the period was over. Moreover, these districts were often subject to famines, and this affected most of the workers from the lowest classes and castes. Coffee planters turned to this source of labour, and were able to offer sufficient monetary incentives for workers to migrate to the coffee plantations (SLNA 1841).

This early migration was more or less spontaneous, and continued on a temporary basis; the workers, largely male, coming over for the harvesting period which peaked in November, and returning in a few months (SLNA 1874). However, as the cultivation of coffee expanded in the 1840s and 1850s, the planters' need to regularize and expand their labour force resulted in the creation of the Immigrant Labour Commission, in 1858 (see ILC Report 1859), for the specific purpose of 'encouraging and improving the immigration of coolies from the south of India'. During the years 1843 to 1880, a total of 275,418 Indian workers entered the country, and 187,062 returned home, leaving a residual population of 88,356. Between the years 1891 and 1930 a total of 468,132 came and 320,744 left, leaving a residual population of 147,388 workers. At the same time, there was an increase in the sex ratio, as migration took on a familiar pattern. While female labour was only 2.6 per cent of the total labour force in 1843, this had increased to 26.9 per cent in 1866 (Ferguson 1866–8: 183 and appendix 2). From then on, this proportion was to increase steadily until women constituted more than half of the total labour force (DLS 1977: 12). This was something that was actively supported by the planters and the Immigrant Labour Commission, who were pre-occupied with getting a cheap and stable labour force. Women, in the Indian situation, were considered to be more patient than the men, and could also be paid less (SLNA 1860).

The Plantation Tamil Community and Labour Control

In time, what was to develop on the estates was a distinct 'plantation community' characterized by features common to many other plantation systems. According to one author, a plantation community constitutes a variety of sub-groups, structured into a hierarchy. This hierarchization is based on 'differences in status and differences in income, wealth and social power' (Padilla 1957: 28). Restricting the analysis to the series of relationships within the bounds of the estate, we will now examine the implications they held for the Tamil community, in general, and the Tamil women, in particular.

Recruitment

Labour from South India was brought across to work on the plantations in what came to be called the *Kangani* System. The *kangani* was a recruiter, usually from the area, who was given sums of money in exchange for the labour he could supply to the plantations. Because of his relatively higher caste status and power, as a recruiter, the *kangani* was able to command respect. The workers were, in this way, divided into a number of sub-groups, each under its own patriarch, or *silara kangani*. These sub-groups, in turn, were generally caste homogeneous, and comprised kin-groups. The whole workforce was structured in such a way that these divisions enhanced what was seen as the 'family principle' (LC Report 1908).

Workers, in family groups at each level, exercised rights over those below them, and paid respect to those above them in the hierarchy. Such methods were said to produce the most 'satisfactory results', and the workers in these circumstances were usually said to 'settle on the estates' (CLC Handbook 1935: 20). The *kangani* was de facto leader of this workforce; he recruited them, travelled with them and controlled them, not only during the journey, but also on the estate. He also acted as the intermediary between the management and the labour force as a whole.

Labour Control

Sri Lankan plantation labour was incorporated into a form of production whose management used race, colour and social prejudices to control the workforce and extract the maximum possible surplus. The head of the organization was the planter, a European who was held in supreme respect by all others on the estate. Below him were his assistants, who were, especially in the early days, white and European as well. Next in line were the office staff – accountants and book-keepers – who, in many cases, were Tamil with some English education. These were not associated with the Tamil labour force, and looked down on them. At every level there was fear and respect for superiors in the hierarchy. Further, European and white supremacy was ideologically inculcated, and this was a significant factor in the plantation management and control system. To this extent, there were similarities with the experience of Latin America and the Caribbean where the 'darker races' were viewed as 'inferior', and where social controls included those based on the privileged position of 'white' and European. The colonial government and state, with their active support for the planters, provided the necessary economic and political base needed to bolster this hierarchical system which, in turn, helped perpetuate this system of labour control.

At present the hierarchical order is manifest in the residential patterns of the different groups. The planters, generally known as the superintendent or *Periya Dorai*, and his family live in a bungalow – large and beautiful – usually with a spacious garden and a scenic view. Depending on the needs of the household, several workers are employed there on a full-time basis. The assistant superintendents usually have other facilities on a slightly lower scale of luxury. The staff are housed in quarters, usually single or twin cottages of a modest size, and situated separate from the rest of the estate labour force. The workers live in 'line' rooms, barrack-like structures with one room allotted to each family. In the early days of the tea industry in Sri Lanka, these lines were allocated, taking caste considerations into account. Today the situation does not allow much room for such adjustments, as the predominant feature of these 'lines' is lack of adequate space, leading to serious problems of overcrowding.

Charles Kemp (1985) argues that the whole system of hierarchy has been held together by what he calls the 'plantation ideology'. According to him, there is a 'belief' or 'acceptance' in the 'self-contained world of the estates' which lay at the core of the plantation ideology. The spatial reality and ideological elements serve to devise an effective pyramid of hierarchy which forms the basis of labour control on the plantations. Within the realm of the estate, authority is virtually unquestioned, as respect for those who were 'higher' is the accepted way of life. In the same vein, integration into estate life involves a commitment to the pattern of segregation and spatial and social divisions as this is ideal to the work and life-style in this environment. It is important to analyse how these principles, apparent at the level of ideology and environment, supplement and are compatible with the operational aspects of production and reproduction on the estate. In this process, what would become clear is the connection between the overall features of control and the way this is concretized in the day-to-day work of the labour force.

Estate Tasks and their Labour Requirements

The plantation crops tea, rubber and coconut have different labour requirements which vary seasonally and, to some extent, over the years. However, it is possible to have an approximate idea of what is considered adequate. The oft quoted figures indicate that employment per hectare of tea ranged between 2.1 to 3.7 labour units, the equivalent figure for rubber being 1.2 labour units per hectare. Coconut had the lowest labour utilization rate of some 0.25 labour units per hectare. After 1977 the Ministry of Plan Implementation gave a directive to increase the level of labour absorption to 3.71 units per hectare in tea and 1.85 to 2.47 labour units per hectare in rubber. However, in reality, this has resulted in many cases in a situation of redundant labour.

The most labour-intensive task on the tea estate, and one which is of highest importance, is the plucking of the flush, or the immature leaf that appears on the bush. This is a skilled task, and it is this plucked leaf that is subsequently sent to the factories for the manufacture of tea. Plucking is generally done by women who are recruited as young girls, and who have learnt the skills from their mothers. There is a definite preference on the part of the estate management to employ girls who have been resident on the estate, and this is usually justified on the grounds that they can more easily pick up the skill of plucking as they have watched their mothers, and often helped them, pluck the tea. Young girls usually start plucking for wages from the age of ten or twelve, and their capacity for plucking increases as they get older, becoming in most cases, 'class' pluckers between twenty and thirty years of age. Their fingers must become very supple for efficient plucking, and it is only with practice that this is improved. Some of the women grow their thumb nails as this is supposed to help them pluck the leaf faster and more easily. The management usually knows the best pluckers, who are put to work on the high yielding bushes (VP tea) while the less skilled (and in some cases the men) are put to pluck the seedling tea. As well as being labour-intensive plucking requires both skills and patience. The plucking is usually done in a plucking 'gang' which is under the supervision of a *kangani* (generally a man). It is the duty of the *kangani* to maintain the standard of the plucking and to maximize the quantity of good leaf. This usually involves fairly strict overseeing of the gang, both in the field and during the time of the weighing.

Tipping of the tea bush is another important labour-intensive task on the plantations, and it is done after the first flush has formed on the bush after the pruning, usually a month later. The new shoots should be some 4 to 6 inches above the wood when it is tipped with the tipping-knife to maintain this height. This task is done by women, selected from among the pluckers, who are then taught the method of tipping. Working in a 'tipping gang', the women are supervised by the 'tipping *kangani*' (again usually a man). It is important to note that the jobs in which women are concentrated, i.e. plucking and tipping, are time jobs. The jobs done by the male workers on the estate are generally piece-rate jobs. Pruning is done by men and it is a task job. This works out to approximately 150 bushes of old tea a day, a task which can be completed in about five and a half hours.

It is clear that the techniques of production are, by and large, labour-intensive, and few, if any, tools are required for the main tasks involved in the cultivation of tea. The skills required for these tasks are learnt on the job, and a preference is given to those who have been associated with the estate. There is a high degree of supervision at every level of the tasks in order to check that the workers are doing their jobs properly. The hierarchical principle, mentioned earlier, is illustrated at the level of the tasks, where the gang is clearly under the authority of the *kangani*. There is a clear domination of women by men in this hierarchical structure. The notion of female inferiority has far-reaching implications for the position of women on the estate, resulting in male dominating structures at every level, giving rise to constant male supervision over their activities as well as a devaluation of their material contributions to the production and reproduction process on the estates.

The Household Unit of Production and Reproduction on the Estates

It is important to take into account the special role of the household on the plantations and analyse the work done, within this unit, as part of the labour process on the estate. This is related to the specific nature of the plantation where its enclave structure and other characteristics typically associated with a total institution, integrate all work done within its geographical boundary into the overall rationale of this unit of production and reproduction. In a very real sense, therefore, the analysis of the labour process would also include an examination of the nature of the work and the power relations within the household.

The three features of household labour on the plantations that warrant such an inclusion are:

a) The fact that most of the labour resides within the boundaries of the estate. There is a certain intermittency, in the jobs undertaken on the field and within the household, which creates a degree of continuum in the range of activities to be undertaken; and which underlines the close relationship that exists, both spatially and in time, between the two areas of work. In effect, it is possible to overlook this internal separation, and view the different tasks as part and parcel of the same structure of work.
b) There is a marked similarity in the nature of the tasks undertaken. They tend to be labour-intensive, time-consuming and monotonous, and perpetuate the same sexual division of labour.
c) Finally, there is a clear connection, related to the above two points, in aspects of labour control in these two spheres of work. This is of particular significance as far as women are concerned. Male domination in one sphere, both shapes and reinforces the same tendency in the other sphere, and places women in a rather extreme form of subordination in the overall structure of work. The clearest illustration of the reinforcing nature of labour control becomes evident when the daily life of the estate woman worker is examined. Being at the bottom of the hierarchical ladder, her position, on the field and in the household, brings into stark reality the way in which her role in the household is an integral part of the labour process on the estates. We have chosen for our purpose the situation of the female tea plucker, as that is where the women workers are largely concentrated.

Women are employed on tea estates, primarily for the labour-intensive task of plucking. Few do weeding or take part in the sifting operations in the factory. However, the plucking, and the associated task, tipping, are the main realms of female work. Women tappers are also important in the rubber plantations. Few workers are employed on the coconut plantations but here we find women in the weeding operations.

The daily life of a tea plucker starts before sunrise. For the villager, who has to walk long distances to the field, this could mean getting up at 3.00 a.m. The resident tea plucker gets up by 4.00 a.m. These women then have to fetch water (and on most estates this involves walking a considerable distance). They then tidy up the house and set about the task of preparing the meals. Given the tight schedule for the day, many women make the mid-day meal in the morning. Non-residents make their noon meals to take with them, as in most cases distances are too far for them to return. In any case, they first serve the meal to the men in the family, and then to the children. After this they prepare the children for school. If they have small children who do not go to school they will have to prepare milk for them. They then have their own meals, tidy up the 'line' room, and send the children to school. In the case of the non-residents it is too early for them to send children to school at the time they leave, so this job has to be done by some other member of the household. After that they leave for work, on the way, dropping off their smaller children with the creche attendant, if there is a creche; or with the *ayah*, whose task it is to look after them. If the creche does not give free milk, she gives the prepared milk to the attendant at the same time. All this has to be completed before she reports for muster at 7.00 a.m., or on days when there is cash plucking, at 6.00 a.m.

Clearly, in view of the number of tasks that have to be completed, some are often left undone, or are done badly due to the pressure of time. When she reports for muster, she is told in which field she will be working. If by any chance she is late reporting for work, sometimes by even a few minutes, she will be unable to work until the following day. She will be 'chased away' by the *kangani* or by the field officer in charge. In this case, she will not be paid, and will not be entitled to any earnings for that day. This attitude is sometimes moderated if the woman is pregnant, in which case the time lost is made up by extra work-time at the end of the day. The woman collects her basket from the muster shed and proceeds to her place of work. This, in itself, might take a considerable time, as the distances involved on large estates can be several miles. Once she has reached the fields she plucks continuously until 12.00 noon. At noon, she walks with her load to the weighing shed where she waits until the load is recorded, and she is temporarily free from field work. She then goes to collect her children from the creche (*pulekamer*) and returns to the lines. She prepares the food for the family, serves the men and feeds the small children. She then eats herself, often feeding the babies at the same time. If she has the time, she washes the dishes, tidies the home and sets out again for work. On the way, she drops the younger children and babies at the creche, and she has to report for work in the field by 1.00 p.m.

The afternoon is devoted to the same work as the morning, i.e. the plucking of the tea leaves and bud. If some irregularity in plucking is committed, such as plucking a hard leaf and slipping it into the basket, the *kangani* reports the fact, often 'blackguards' the plucker and reprimands her for her carelessness. If the matter is taken up with the senior officers, she may be sent home. In the majority of cases she is either strictly reprimanded or fined in some way. The presence of mature leaves, if detected, could lead to punishment for the overseer who had overlooked it and, if only for this reason, they (and they are invariably men) are particularly careful in supervision. It has often been remarked that 'this sort of strict supervision compels the plucker to bring in good leaves'. On the days when there is a normal or poor crop, the work usually stops at 4.00 p.m. On days when there is an abundant crop, the women may pluck much longer, sometimes until as late as 5.30 p.m. After that, again, they have to carry their leaves to the weighing shed or some central spot for weighing. The weight recorded is noted down on 'chits' (small scraps of paper) and in the check-roll book, and the women return once more to their homes. In the cases of the residents this would normally be between 5.30 and 6.00 p.m. In the case of non-residents, much later, the women reaching their village homes at about 7.00 p.m.

The weighing procedure is a point that warrants a small note by way of elaboration. Weighing is carried out at least twice a day, and it is done 'outside' the working day. This means that it is outside the eight hours of work done in the plucking field. There is a weighing area, sometimes near the field, but often a considerable distance away, to which the women take their baskets of leaves. If there are three weightings, the first is at 9.00 a.m., the second at 12.00 noon, and the third at 4.00 p.m. At the weighing shed the women stand in queue, and some time is spent while they look through the leaves to locate coarse ones or twigs which may have accidentally fallen into the basket. When this is done, the leaves are weighed and the weight noted down by the officer in charge. One kilogram is normally taken away for the weight of the basket, and in the rainy season up to two kilograms are subtracted for the moisture that is in the leaves. The weight is written down on a small slip of paper (the chit) at the early weighings and this is pinned on the woman's blouse, or put into her basket. Sometimes nursing mothers are allowed time off after the first weighing to feed their babies. One hour is normally allocated for this purpose, but there are still estates which do not recognize feeding time. When they do, the women go to the creche, if it is nearby, or some older child arrives with the baby, and the mother sits down by the patch and feeds her child. Soon afterwards, she returns to the plucking of the leaves. After the final weighing, the officer in charge collects the various chits from her and records, in a check-roll register, the total weight of the leaves brought in. Most workers maintain that on many estates the picking norm is so high that few of them can reach it, and that, anyway, they receive only 17 cents per kilo for 'over-weight' and 25 cents per kilo in times of cash plucking.[1]

On reaching their homes, having collected their children again on the way, they immediately set about various household chores. They collect the firewood and fetch the water needed for the preparation of the evening meal. Men and children sometimes help in collecting the wood. The women bathe the children and then clean themselves. After that they wash the afternoon dishes, if they have not been done, and set about preparing the evening meal. In the case of non-residents, the rush to do all this is even greater because they arrive back at their homes at a much later hour. After the meal is prepared, the women serve their husbands and the other men of the household. Then the children are fed, and it is only when that is done that the women have time to sit down and to eat their food. Next they prepare the children for the night. In the case of residents, the women eventually go to sleep at about 10.00 p.m., while in the case of the non-residents it is normally later. There is very little difference, in any day, in the pattern of life of the female tea plucker. Whether it is raining or whether it is not, she has to follow the same routine of work on all work days. The length of the work day may vary a little with the nature of the crop. When there is a good crop, she may be working in the field from 6.00 a.m. to 5.30 p.m.; but in any case, in spite of this, she is responsible, and she herself has to undertake all the other tasks that have been specified above. There may exist exceptions to this pattern, but they are few and very far between.

Household Tasks on the Estate

The most striking feature of household tasks on the estate is the maintenance, whenever it is physically possible, of the division of labour that existed in traditional Tamil society; and this is true regardless of caste, economic status or place of residence. It is based on a traditional system in which the man was primarily responsible for activities outside the home, earning the money, doing the shopping, and generally bringing home the means by which the family members had to live. The woman's role lay in cooking, cleaning, caring for the children, fetching water and firewood and, in general, catering to the needs and services of her husband and children. Women helped their husbands in the field during the peak seasons (again with a strict division of labour in which work with the plough and the *mammoty*[2] was strictly reserved for men), but it was unheard of for men, of any caste or occupation, to assist in home chores. In the case of estate labour, although the women earn and also work all day in the fields, their tasks within the household remain the same. Change in the production system has had little effect on the division of labour in household activities, except in a few cases where this extra workload is such that the traditional pattern is physically impossible.

The attitude of the estate management and of the society, in general, to household tasks rationalizes the double or multiple workload of women. Firstly, these are not considered as 'work' but are seen as, in some way, a 'natural' function of the woman. They are taken for granted, and no thought is given to the extra burden that is being imposed. Secondly, when the household is faced with an economic crisis, it is mainly the woman who bears the brunt of it because work, in the household, subsidizes the fall in income, up to a point. Finally, because the man and the children are given priority, and generally secure the most and the better food, the woman's general health and nutritional standards deteriorate most. The additional strain and effort on her part go largely unseen because the entire system views the household sector as essentially 'non-work'. She is always working, and additional tasks are readily placed on her in the home because the prevailing ideology does not recognize them as 'work' at all. This is, consequently, yet another illustration of the way in which women are marginalized in the estate community. Women's marginalization can be better understood in the light of traditional notions about women and their role in Tamil society. For this purpose we need to discuss the notions of caste hierarchy and sanskritization.

Caste Hierarchy and Sanskritization

Caste Hierarchy on Sri Lankan Estates

The two main divisions found on the Sri Lankan estates were the *Sudra* and the *Adi-Dravida*. In order of hierarchy among the *Sudra* were two sub-divisions, namely *Kudian* and Non-*Kudian*. A common caste found on the estates was from the *Kudian* sub-division, known as the Vellalan who were generally accepted to belong to the highest social caste on the estates. Many of them, in fact, claimed to belong to the *Vaisya* division. Besides the *Vellalan* at least thirteen other castes and their further divisions belonging to the *Kudian* and non-*Kudian* sub-divisions are found on the estates.

The *Adi-Dravida* or *Panchama* formed the largest proportion of the estate workers. In Tamil Nadu in South India they, historically, worked as some form of hired labour, and they may even have had to scavenge the area. The *Chakkiliyan*, considered to be the lowest caste (their name is derived from the Sanskrit word *Shatkuli* i.e. 'flesheater'), were often given the task of sweeping (apart from the usual field work for estate production). The *Pallan* and the *Paraiyan* clearly constituted the bulk of the labour force on the estates. These two castes in Tamil Nadu worked for the *Vellalan* and the other high castes, and often had their dwelling places outside the village areas. Given their social situation, it is perhaps not surprising that a large number of converts to Christianity came from these two castes, especially the *Paraiyan*.

Sanskritization Process

As the labourers from South India were incorporated into the plantation system, they were to experience a change or a shift in their status as workers and as members of the new estate community of which they were a part. Their new status depended not only on the particular caste from which they came, but also on the caste ranking in the new environment, and on whether they were male or female. In general, most of the sub-castes experienced upward mobility as the process of migration freed them from the controls of the higher castes. At the same time, however, while this implied greater freedom and power to the male members, it created greater controls on the women. An important element in this process which continued to control their lives was the retention of some caste values. This was facilitated by the peculiar pattern of recruitment which encouraged migration in kin-groups under the leadership of a *kangani*, and did not allow the workers to free themselves completely from their caste background. Furthermore, since they came across in such groups, it was easy enough for the relatively upper-caste groups to demand the services and respect which were their due previously in the Indian situation.

Even the relatively higher-caste Tamils who came across as estate workers were, in reality, from the relatively lower rungs of the caste ladder in the Indian context. Understandably, they were less concerned about strict adherence to the rituals, religious ceremonies and customs that were part of the society of the upper castes; if only because these were often seen as the domain and right of the latter. There were cases to indicate that compared to the caste rigidity that characterized norms and behaviour of the upper castes, there was relative flexibility of caste divisions among the *Sudra*.

As far as *Adi-Dravida* women were concerned, their profession as cultivators or scavengers had forced them to work together with the men. They had much contact with workers in their own sub-caste, and often even with the members of other sub-castes. This situation created relatively independent attitudes, as well as a bolder and more egalitarian attitude vis-à-vis men of the same sub-caste. They often did not adhere to the strict moral codes of the upper-castes which placed women under the control of their men-folk, and it was not uncommon for these women to enjoy a greater degree of sexual freedom. Inter-caste marriages were more widely accepted, and in the case of the *Adi-Dravida*, all of these trends were more pronounced. In spite of these factors, things were to move in a more conservative direction for women in their new situation as plantation workers in Sri Lanka.

There was a very conscious attempt by the higher sub-castes among the *Sudra* to (re)introduce a similar pyramid in the new environment. This was a conservative and traditionally-based caste structure with themselves at the top, taking over the role of the *Brahmin* and the other high castes. This attempted reconstruction of the former social background, with its norms and behaviour patterns resulted in the imposition, in this setting, of a quasi caste system. The hierarchy was now one with the *Sudra* (often claiming to be *Vaisya*) on top and the *Adi-Dravida* below, with various sub-castes being highly differentiated among the *Sudra*. The demarcation between the different sub-divisions grew more significant and more important as those at the top of the pyramid separated themselves from those below. Caste ideologies and patterns of behaviour which gave them a higher position within their community were revived.

In particular, the method of recruitment, the system of labour management on the estates, and their relationship to the wide society in Sri Lanka stimulated this process. Firstly, the fact that migration took place under the supervision of a higher caste *kangani* and usually in families or kin groups together, served to make the preservation of caste identity much easier. Secondly, when these groups arrived on the estate they were isolated from and had little social contact with Sri Lankan society. The latter had, anyway, a social system which was feudal in nature, and inimical to that of the Indians (de Silva 1979: 129–52 and Silva 1979: 43–70). Thirdly, the formation of labour gangs on the estate, under *kangani* and sub-*kangani*, was (at least in the early days) based on caste, and tended to strengthen both caste and kinship ties. It has been argued that the *kangani* system in fact tended to perpetuate these values on the estates. Given this tendency for the sub-*kangani* groups to be relatively homogeneous in terms of caste and kinship, and given the fact that, in the earlier days, these groups had to align together under a head-*kangani*, a situation was created in which, in reality, a large number of *Adi-Dravida* had to work under *Sudra kangani* (Jayaraman 1975: 57–64). In fact, the *kangani* used caste in this way to maintain his position, assert his authority, and better his situation. He could control the workers below him, and this gave him enormous power in negotiating with the management. Caste differences between the workers were expressed in a number of ways in their social life. Living quarters, for example, were arranged on the basis of caste. Higher-caste people disliked living in the same lines as people of lower castes, and in some cases objected even to having to look upon them. These 'arrangements' were nearly always made on the advice of the head *kangani* who knew the 'needs' of the different castes, and allotted rooms in line with such reasoning (Green 1925). The result was often a concentration of a particular caste in one 'line' or one housing area, keeping awareness of caste very much in the fore.[3]

In these new circumstances hypergamous marriages, especially among the higher castes on the estates, were strongly condemned, and if the castes were not structurally close, this could result in virtual ostracism. All sub-castes of the *Vellalan* were discouraged from inter-marriage (although exceptions were not uncommon), and the *Konga* sub-caste (the lowest amongst the *Vellalan* caste) was very often not allowed to mix socially with the higher *Vellalan* sub-groups (Green 1925). The significance of this process lay in the increased differentiation that developed amongst the workforce on the basis of caste.

Religion played a very important role in the lives of the estate workers; and rituals and rites, formerly the domain of the *Brahmin* and other higher castes, were now taken over and more rigidly observed by the *Sudra*. There was also a caste hierarchization experienced in these practices, with the *Adi-Dravida* not being allowed the same privileges as the other higher castes. If an *Adi-Dravida* cooked the food, the *Sudra* would refuse to eat it. In these many ways the divisions within the Tamil community were heightened in their new setting.

The Impact on Female Labour

The above factors had a special significance in terms of the position of women. The resurgence of caste and religious practices gave prominence to traditional notions concerning women and concerning their role in Tamil society. Most important of all these was that women were viewed as inferior to men. It was their duty to serve the needs of men and to be subordinate to them. Inequality between men and women, an important feature of traditional Hindu practices, became even more apparent in the lives of the Tamil workers on the estates as they aspired to positions of status within their community. Subordination of women to men was considered 'proper'. It could be widely used to control their lives (sexually and in terms of other services), and these beliefs were constantly being reinforced in the social life and took place on the estates.

Festivals were, in general, important events, and they were even more so in the lives of women because these were important opportunities which allowed for wider social contacts outside their work situation and their immediate neighbourhood.[4] Religion of higher castes of non-Brahminic order, became a more important part of their lives; and the inequalities inherent in this religion became an even stronger means of male control. Women could not set foot inside the temple's sacred area 'lest their action pollute the gods'. The important ceremonies following births were much more elaborate in the case of male than female children. A male child was always considered to be that much better. When girls reached puberty, this too was the subject of ceremony, but in the course of their menstruation they were confined to a separate section of the house, as they could 'pollute' other family members.[5]

The marriage ceremony emphasized the subservience of women;[6] and following a death, it was the men, and not the women, who accompanied the deceased to the graveyard.[7] All these points served to engrain a sense of their inferiority into the women and into the consciousness of the Tamil community. These rituals were rarely observed with such sex or caste segregation among the *Sudra* and *Adi-Dravida* who lived in Tamil Nadu. The process of sanskritization, as such, created a loss of freedom for the woman. As a *Sudra* family tried to emulate the behaviour of higher castes, and reflect the status that it had now acquired, the woman in this caste assumed the role of being the private property of her husband and subservient to him. At the same time, this process also implied distinguishing oneself from other people, or a process of social differentiation, as discussed earlier, which resulted in the woman becoming more isolated and even more cut off, at a social level, from the lower-caste women and the rest of the estate community. Thus, with the sanskritization process emphasizing caste differences, *Sudra* sub-caste women began to lose the freedom they had earlier enjoyed. This was manifested in *Sudra* women by their special efforts to avoid alcohol, and by an increasingly conscious observation of rituals associated with high-caste festivals. Women of the lower divisions of the *Adi-Dravida* castes often drank. Their attitude in general was more relaxed, and they had less to gain from adopting, and were therefore less concerned with the need to follow, high-caste rituals. This was marked in the attitude that they displayed to men in their households. Even today, although all women claim to hold their husbands in fear or *bhayam*, there is a correlation between the level in the caste hierarchy and the degree of reticence that women display towards their men. The higher the caste, the greater the degree of reticence that they show. *Sudra* women tend to be less free and more subservient to their husbands than women amongst the *Pallan, Paraiyan* and *Chakkiliyan*. The latter, in contrast, are more open with their men and are often prepared to stand up and fight for their rights.

The implications of this entire process, in terms of labour control, cannot be underestimated. It was largely in the interests of the management to promote social relations which upheld notions of respect for those in power, and which discouraged any type of joint organization or confrontation on the part of labour. In this way, using the caste values, the plantation management could rely on obedience of the workforce to the *kangani*, and at the same time keep them sufficiently divided. With regard to women, there was also a special significance, and one which allowed greater economic exploitation of labour. Firstly, the notion that women were inferior to men justified their being paid lower wages, even when they did the same work. Secondly, the division of labour was such that women were involved in monotonous time-consuming activities which were labour-intensive and of lower status, although this was justified on the grounds that women were more 'patient'. Finally, the profitability of the estate was also promoted by the structure of the household unit and the division of labour within it. These activities which included cooking, cleaning, looking after the children, were almost exclusively done by the women and seen as part of the 'natural' duty of women, and as such did not have to be accounted for by the management. In this way, patriarchy and casteism were to become important elements of the Tamil community; and used as a means of labour control for women on the plantations.

Inroads made into the Forms of Labour Control – the Contemporary Situation

There have been two major processes that have served to challenge the traditional patterns of labour control. Their impact has been limited and in some cases even marginal. However, it is important to explore this a little more. The first process has to do with the granting of universal franchise and the subsequent development of trade unionism on the plantations. The second is concerned with the repatriation process through which large numbers of plantation labourers have been rendered 'stateless' or been returned to India.

The granting of universal franchise in 1931 under the Donoughmore Constitution served to question the traditional basis of authority in the plantation Tamil community. The most important effect of this was to shift the balance of political power to those sections of labour that were most numerous, notably the *Paraiyan* and the *Pallan*. The high-caste workers who had hitherto maintained their superior position with the caste rationale now had to give way to a situation where the most numerous low-caste groups had a more important influence on political organization. However, although efforts may have been made to shift away from power structures based on caste, it had been so much part of the social situation that in one way or the other, even the trade union movement was affected by this phenomenon. Although this is denied by the existing unions, and efforts are made to erode power based on this, it nevertheless continues to have importance in the social life of the workers.

The second element to question this traditional caste authority has been the increased process of repatriation in the post-independence period. Many of the people who had opted to return to India in the wake of the Sirimavo–Shastri Pact of 1964 and the Sirimavo–Gandhi Agreement of 1974 have been people from the higher castes. These people had, in many cases, maintained contact with relatives in India. They had often made journeys to their native villages, and they were even in the habit of remitting money to their families in India. It was the habit of persons in those castes to save their money, while the lower castes, with few possibilities for their future, were more likely to spend it on alcoholic drinks or speculation. When the process of repatriation was started, those families with some contacts in India initially opted to leave Sri Lanka. The net outcome of this is that there are few such families remaining on the estates, thereby making it possible to break down the values that they had stood for.

Finally, there is the increasing social intercourse that has come about between the Sinhalese villagers and the Tamil workers, this phenomenon being more evident in the mid and low country areas. In some cases this has given rise to inter-marriage and, in the process, a more direct questioning of the traditional value system.

In all these cases, however, there has been little to question the traditional position of women. It is true that women in the lower castes tend to be more vocal about their problems than those of the higher castes, and to that extent it might seem that their grievances are given more attention. However, with few exceptions, the trade unions have tended so far to be male-dominated. The participation of these women is more or less limited to paying their membership fees and making their complaints to the *thalavar* (union representative). This is further compounded by the fact that meetings are often held when the women cannot be free from work (either in the household or in the field) and little effort is made to provide facilities whereby their jobs can be taken care of. Even when they do attend, there is a tendency for the meetings to be male-dominated. It is true that there is concern amongst some of the younger generation who are more active in asserting their rights as workers. However, this is more the exception than the rule, and most of the women continue to feel that trade unionism is a man's domain, and therefore do not take part in the decision-making process.

Social Practices and the Law

Many of these social practices have a certain bearing on the way in which legal and other formal norms are expressed and implemented. The most glaring feature of the legal discrimination that women workers have experienced is the inferior minimum wages they have received for the same work as men. These wages are not haphazard or disorganized, and there are no variations from one area or part of the country to another. They are fixed and governed by a Wages Board to which the government appoints members, on which trade unions and employers are also represented, and whose task it has been to establish the minimum wage since 1944. Although the government, after the strikes in April 1984, decided to redress this inequity, the women workers had been receiving minimum wages that were some 25 per cent lower than those of men. This was further compounded by the fact that the supplements they received were based on a cost-of-living index which was supposed to account for the changing cost of living. The minimum wage is multiplied by this index, and since men had a larger initial wage, the supplement they received was consequently larger. This discrimination, as we have seen, had its historical roots, and was further strengthened by the value system in the society and by the production system as well.

Another important factor which seems to have been removed, after the April 1984 strike, concerned the practice of paying the workers once a month in accordance with the number of days they had worked, usually only those when work was offered, which varied according to the crop, the area, the season and the yield of the particular estate. So that although the minimum daily wage was fixed, many workers experienced great hardship, and were forced to incur debt during the lean months. The law, in 1974, provided that the workers were entitled to 108 days of work every six months, but this did not necessarily mean that eighteen days' work was guaranteed every month. In 1984, the government affirmed that in accordance with the Estate Labour (Indian) Ordinance of 1887, a minimum of six work days per week would be offered. However, the actual implementation of this law in the subsequent period still has to be studied. It must be noted that women, in particular, faced the hardships of the previous situation, where because of the structure of the wages, the division of labour and the lack of mobility experienced, the variations in income for women were larger than those for men, as studies have shown (Kurian 1982). Further, it has been shown that women worked longer hours and still received less income from their estate work (ibid). So long as the division of labour continues to maintain the women in labour-intensive, time-consuming activities they will continue to work longer hours than men. The tea plucking, for example, takes the whole working day, while the men, on a task, generally finish their work by about 2.00 o'clock in the afternoon.

Women were also far from being compensated proportionally for any extra efforts which they made in trying to make up for a shortfall in monthly income. For example, tea pluckers were required to pick a daily norm of so many kilograms. If the daily minimum rate was divided by this norm it could be expressed as a rate per kilogram plucked. When women brought in more than this norm, they were paid for 'overkilos', at a basic rate of Rs. 0.17 per kilogram. This was raised to Rs. 0.25 if they 'turned out' for more than 80 per cent of the days on which work was offered. The point was, however, that in general, the rate per overkilo worked out to be 30 to 40 per cent lower than the rate per kilogram during the normal day.

More than that, women pluckers received a declining rate for each additional kilogram they brought in over the norm. In other words, women who were struggling for additional income through greater effort were not only paid for this at a lower rate, but at a rate that became lower and lower the more work they put in. Further, the rate of decline in payment for extra work was faster in the lean months when the crop had a low norm. Those months in which they had difficulty in reaching the norm were also those in which the payment per kilogram fell faster. In the very months when they had to work harder to reach the basic norm, and when it was more difficult for them to pluck any overkilos; if they actually did so, the rate of payment would decline more sharply. In short, the women who were forced to work for additional income were the women being exploited more. A similar situation prevailed on the rubber estates.

An area in which the ramifications of labour needs and social practice actually lead to contravention of the law, is in the payment and receipt of estate income. Although there are some exceptions, the common pattern on estates is for a male member (normally husband, elder brother, father) to collect the wages for all the working members of the family. This is accepted by the men (and women in most cases) and is justified at different levels. Firstly, women have longer hours of work in the field and in the household, and cannot spare the time to collect the wages. Secondly, the management and the society accept the elder male as the head of the household, with all the rights over the income of its members. Thirdly, the man is largely responsible for the shopping and contracts outside this unit (be it with the moneylender or even the management) and he would be spending this money anyway. And finally (and this is also reflected at some trade union levels) the poor education and illiteracy levels of women make them more vulnerable to duplicity and cheating. For many such reasons, it is not unknown for the money earned by the women never to reach their hands.

While this division of tasks might be thought to be a harmonious situation, it must be noted that it is not always so. Women are known to complain about the way men spend a large proportion of their income on liquor; and physical fights on pay-day are not uncommon on this issue. There was one instance when the option of the women regarding the payment of wages was opposite to that put forward by the trade unions. In this particular case, the trade unions had demanded that estates management not deduct the cost of the rice ration from the pay packet as they were dissatisfied with the quality of the rice, and would prefer to buy it from outside, and not through the estate as was the normal practice. Before the management reached a decision, they consulted with the men and women workers, and found that the women actually objected to this change. Their argument was that this deduction from the wages for the rice meant some element of guarantee about the food, which they would not have if all the money was given, as wages, to the men.

The area of welfare facilities also reflects the controlled circumstances of the women's lives. Here, as in the case of social practices, the conveniences provided cater largely to the labour needs of the plantation management and to the general pattern of patriarchy. Creche facilities, for instance, are totally inadequate, and exist, to a large extent, in order that the mother can put her children in some place, and thereby be available for work. While a few creches have been upgraded in recent years, the tendency to employ uninterested non-Tamil-speaking creche attendants (who have the job through some political influence with the local member of parliament) has proved to be a disincentive to these Tamil women; and has created other problems for the women and their children. Medical facilities are, by and large, of a curative nature, and the unhygienic local conditions of life are seldom taken care of. While the maternity service, where there are hospitals, is adequate, it is not uncommon for the maternity benefits to be given to the husband, and to be used to pay off accumulated debts. In fact, it is not unknown for the moneylender to give additional money on the strength of a woman's pregnancy. In this way the benefits due to the women do not reach them.

The housing situation is another area of importance to the woman as she spends a considerable part of her non-estate work within it. Legally, a single 'line' room may not house more than two adults and three children under twelve years of age. A double room cottage should not accommodate more than four adults and four children under twelve years of age. These instructions are laid out under the Disease (Labourers) Ordinance for 1961 (Section 12). While these requirements themselves may seem inadequate, it is clear that there are many cases when even this is not adhered to, and several generations of families are known to live together.

Education is another area of gross injustice to women, and their high rate of illiteracy testifies to this. The illiteracy rate of nearly 90 per cent is a reflection of the two elements noted earlier – the labour requirements of the plantation and the value system of the society. Women were seen, and still are, to have no other future than to get married and pluck tea. As such, education was considered both unnecessary and expensive as far as a girl's family was concerned, especially in a situation where she could be plucking tea. As early as the first decade of this century, this was the main argument put forward by the management for not having compulsory education for children. And while this has clearly changed since that time, as far as the males are concerned, we find that the attitude is more tenacious as regards girls. This, in turn, makes the women dependent on the men for many negotiations and even for understanding political issues, thus reinforcing the pattern of control over them.

An area in which the state has been active in recent years is the area of family planning, which in some estates has shown remarkable success. However, while the phenomenon of unmarried mothers is not uncommon, no information is provided for them. Furthermore, when information is available, the stress is on terminating, rather than on controlling fertility; and pressures are brought to bear, encouraging either tuberctomy or vasectomy. The state (through the estate management) pays up to Rs. 500 to a woman to undergo the tuberctomy, compared to Rs. 250 only for a vasectomy in the case of a man. In these situations it is often clear that both the man and the woman are hesitant about taking the operation, and in many cases associate it with pain and ultimately a loss of strength. In the case of men, it also implies a certain fall in manhood and self-esteem. But the higher compensation and the fact that ultimately the woman is responsible for looking after the children make her opt for the operation. As in the case of maternity benefits, the money gained is often used to pay off family debts (and it is not unknown that women have the operation solely for this purpose). In this way, the state and the society are involved in influencing the decision for the woman.

There is one area, however, that neither the society nor the state has acknowledged openly. Nor do they see any need to provide any avenues of redress. This has to do with the violence to which women are subjected on the estates. This abuse, which can take the form of physical threats and harm, has had a long history in plantation agriculture and plantation society, and is not confined to Sri Lanka. It has always been seen as a domestic matter, something private, and something which, with the exception of the most extreme cases, was no business at all of society at large. This attitude has been reinforced by the whole ideology of patriarchy, in which women are essentially subservient to male power structures. There are many instances where women workers on the plantations are used, against their own wishes, for the sexual gratification of men at all levels — the management, the *kangani* and even workers. On contemporary plantations, physical violence between men and women is still an extremely common phenomenon, and is not even commented on as unusual or untoward, except in rare circumstances. Fights and beatings occur, especially after men have been drinking. It is possible for this to happen simply if the food prepared is not tasty enough, or if the woman wakes up too late, or even if she 'talks back' to the man. Fights based on suspicion and jealousy are also not uncommon. Nonetheless, complaints from the society or from the women themselves are rare.

This violence was, and still is, a very real part of the everyday life of the estate women. It is known to occur widely, and is acknowledged even by the management, but, as in many other parts of the world, it is seen as something private. Here in Sri Lanka, no avenues for protection are provided, and violence is often accepted as the right of the man concerned, whether this man be the husband, lover, *kangani* or estate superintendent. Today, if only because of the growing power of estate trade unions, abuse by the management has more or less disappeared, but it is still a prevalent feature of the life of the female worker, and one that has to be seen as yet another element in a long series of controls that are placed upon her on emotional, physical and social planes.

Conclusion

The development of the South-Indian Tamil Community in Sri Lanka was closely connected with the growth of plantations in the nineteenth and early twentieth centuries. Their society in general, and the position of women, in particular, were influenced by the way workers were integrated into this form of production. This was essentially an authoritarian hierarchical structure in which the ideology of white supremacy was inculcated by the management and the colonial state in order to strengthen the power of the planter. Discrimination of another form, i.e. casteism, was also important for the Tamil community in this new environment; and this, together with a strengthening of traditional patriarchy, played a significant role in perpetuating the notion of inferiority of women and their work. This process allowed the plantation management to pay lower wages to women for their work, justified the double workload – field work and housework with no compensation – and gave rise to a situation of male dominance and female dependency in most spheres of women's lives.

Notes

1. Cash plucking refers to plucking done outside the normal workday, and which is paid for by cash, outside the normal income. Usually, this is resorted to by management where there is not enough labour to pluck the flush during the workday.
2. *Mammoty* is a Tamil word. It is a tool for digging the earth, not a spade. It has a much shorter and curved handle.
3. Observed in Green's (1925) book and confirmed in interviews with senior superintendents. They maintained that it was not that 'lines' were consciously allotted on the basis of housing decisions, and that they followed the custom of ensuring that there should be no unrest by placing certain castes in certain areas. The 'lines' refer to

long, barrack-like buildings divided into very small dwelling units, side by side, opening out into a common area.

4. The most elaborate celebrations relate to the festivals of Deepavali, Pongal and Adi Poussai. Also celebrated are Vael, Shiva Ratri, the feast of Perumal and St Anna's Day.

5. The coming of age ceremony of the girl, known as the *ruthu sadangu*, is a significant ceremony in the family, but the confinement of the girl reinforces the notion that there is something unclean about a woman during this period. This feeling is perpetuated by other norms; for example, that she should not touch the food that is to be served to the rest of the family at any time when she is menstruating. In fact, at these times she is treated as somebody who would pollute the others if she came into contact with them.

6. In a typical Vellalan marriage, the bride has to make a circle with the first fingers of each hand and look towards the star Arunditi. This symbolizes wifely duty and constancy. The rituals also stress the subservient role of the female to the male.

7. The funeral rites are considered to be very important on the estate, and only men are allowed to participate in the various rites involved.

6

Indian Migrant Women and Plantation Labour in Nineteenth- and Twentieth-century Jamaica: Gender Perspectives

Verene A. Shepherd

Female labour force participation in the plantation industry of Jamaica and the wider Caribbean has a long history. White indentured servant women worked on the early plantations in the seventeenth century and enslaved African women dominated the plantation field labour in the period of slavery. Indian women were (reluctantly) recruited for plantation labour in Jamaica in the mid-nineteenth century as proprietors searched for a new system of slavery in the aftermath of the abolition of African slavery in 1834, and the subsequent withdrawal of significant numbers of ex-slave workers from the sphere of estate labour. The recruitment of Indian women for commodity production on plantations was not unique to Jamaica as similar patterns of labour recruitment were organized for Fiji, Natal, Mauritius and other territories within the colonized Caribbean (Beall 1991: 89–115; Lal 1983; Laurence 1971: 24–46, 1994; Look Lai 1993: 107–53; Ramachandran 1994: 132–4; Reddock 1984, 1994). Though their numbers were small, Indian women comprised the largest proportion of female immigrant labourers recruited for Jamaican plantations.

The history of plantation labourers of diverse ethnic origins has traditionally been subjected to race and class analysis. The perspective of gender has also now assumed a greater role in the historical discourse of the Caribbean as more and more research reveals that the experience of proletarianization and racial and ethnic oppression was not the same for men and women. In the case of the history of immigrants, as Jo Beall (1991: 89–115) emphasized in her study of indentureship in Natal, a study of the separate experience of Indian women under contract as indentured labourers serves to illustrate how the materialist feminist discourse can point to crucial issues in a broader analysis of relations of gender which can support analyses of class and race and/or ethnic relations.

Accounts of the differential socio-economic positions of minority ethnic groups in late nineteenth- and early twentieth-century Jamaica, for example, usually emphasized the depressed conditions of the Indians. Further examination of the comparative positions of male and female migrant workers usually leads to the conclusion that while it is true that the common history of struggle against migration, racism and the general conditions of indentureship united Indian men and women, as in Natal and elsewhere, Indian female plantation labourers in Jamaica were subjected to ultra-exploitability. Indeed, contrary to Emmer's (1985) conclusions for Suriname that migration and indentureship were vehicles of female Indian emancipation (a claim denied by Hoefte 1987), the indentureship experience in Jamaica did not seem to result in any great degree of social betterment for Indian women, though there were, undeniably, small gains.

This chapter will explore the ways in which gender ideology contributed to the shaping of the social and economic experience of female Indian plantation workers during and after indentureship. Gender clearly played a role in the shaping of socio-economic policies and social consciousness in post-slavery Jamaica. Jamaican slave society had been characterized by a paucity of working-class white women, with enslaved black women comprising the majority of field workers. The gender division of labour, while weakly instituted in the area of field labour, was observed in higher status slave occupations. An ideological shift occurred, however, in the post-slavery period. A combination of factors, among which were European patriarchy and Victorian ideals, imported into the Caribbean, dictated that men should function in the 'public' domain of wage labour while women were to inhabit the sphere of uncompensated work in the home. This process had been taking place in Britain since the end of the eighteenth century. By the time of the Indians' arrival, therefore, attempts were already underway to adhere to a 'proper gender order' in the division of labour. This ideological shift had a significant impact on the recruitment of female plantation workers, explains the marked sexual disparity in migration schemes, and partially accounts for the depressed socio-economic life of the contract and post-migrant labourers.

Gender and the Recruitment of Female Plantation Workers

In order to understand how gender functioned in the experience of Indian plantation workers in Jamaica, it is first necessary to examine how gender affected migration policies and the numbers and types of women recruited. The tendency in the mid-nineteenth century to dichotomize work and family, public and private, determined the landholders' attitude to the recruitment of Indian women. It is clear that they initially regarded the importation of women as uneconomical. In the first place, planters did not regard Indian women as capable agricultural workers. They believed that Indian men worked more efficiently and productively. Their view, as expressed by the Acting Protector of Immigrants, was that, 'indentured women as a rule are not nearly the equal of the men as agricultural labourers', and in the early twentieth century, when efforts were being made to increase the numbers of women shipped, planters objected to being obliged to pay to import women who they claimed were 'not as good' as male agricultural workers (C.O.R.1843–1917a: 571/1). Second, unlike during slavery when black women had the potential to reproduce the labour force (though fertility rates were generally low), the progeny of Indian females could not automatically be indentured; so Indian women were not initially highly valued for their reproductive capacity. Indian children could only be indentured at age sixteen, though in practice many were used in the fields from age six or so, receiving wages of between 3 and 6 pence per day. But this was only with their parents' consent. Furthermore, proprietors were obliged to provide rations for immigrants' children, whether such children had been imported from India or born in the colony. In some cases, they also had to stand the cost of hiring nurses and establishing creches to look after immigrants' young children (ibid). Third, landholders were not too concerned initially about the social life of the immigrants; so the sexual disparity and its implications for the stability of family life did not preoccupy them. Indeed, the requirement to provide immigrants with return passages at the end of their contracts seemingly made it less critical to be concerned about the construction of the Indian family and the impact of a shortage of women.

Planters were thus not particularly anxious to employ indentured women as part of the immigrant plantation labour force. Lal and McNeil (1915: 313) reported that 'so far as we could ascertain, employers are not particularly anxious that women should work provided that they are properly maintained and absence from work does not merely mean exposure to temptation and the possibility of serious trouble'. Some employers were even prepared to reduce women's indentureship from five to three years; and those who did not support this measure only opposed it because of the pressure of Indian men who could not support non-working partners, and planters who argued that women might have too much free time on their hands; that 'a woman who is not occupied otherwise than in cooking her husband's food is more likely to get into mischief'(Lal and McNeil 1915: 313). But even though five-year contracts remained, in the first three years of their contract, as Lal and McNeil observed, 'a woman who is known to be safely and usefully employed at home will not be sent out to the field' (ibid).

Proprietors, therefore, maintained a gender-specific importation policy which favoured men; and recruiters in India mostly carried out the instructions of the Jamaican planters regarding the composition of recruits. Up to 1882, recruiters were even paid less for each female emigrant recruited. The rate paid was 6 annas per head for females and 8 annas for males (IOR 1882: V/24/1210). The need to increase the numbers of women shipped to the region soon caused an increase in this rate. On the ship *Blundell* in 1845 which carried the first group of Indian indentured workers to Jamaica, women comprised just 11 per cent of the total of 261. When the number of girls under age ten is added to this figure, then the percentage of females increases to 15 per cent. On the *Hyderabad* in 1846, women made up 12 per cent of the total shipment of 319 with total females comprising 15 per cent as on the *Blundell*. On the *Success* in 1847, women comprised 10 per cent of the shipment of 223 adults (C.G.F. 1845–1916, Ship's Papers).

Jamaican planters relented and adjusted the unfavourable ratio only in the face of governmental pressure to conform to a 40:100 female-male ratio for immigrants over age ten. But they did not go as far as to support a suggestion by the 1913 investigating team of Lal and McNeil of a further increase in the ratio to 50:100 regardless of the age of the female immigrants. This was despite the support given to the suggestion by the delegates appointed to consider the future of indentureship after the end of the First World War when certain difficulties in importing labourers were anticipated. They recommended that, 'wherever it is possible to find a sufficient number of females willing to emigrate, this ratio [of 40:100] should be increased [to the level suggested by Lal and McNeil]'(C.O.R. 1847–1917a: 571/4).

Planters were also forced to agree to an increase in the importation of women because of the economic imperative of encouraging the settlement of Indians rather than their expensive repatriation. Indeed, by the end of the nineteenth century, planters had successfully influenced changes in immigration policy as it related to the length of contract, repatriation and the period of industrial residence. By that time, contracts had been lengthened to five years, the period of industrial residence extended to ten years after which repatriation could be accessed; and the immigrants were being required to pay a portion of the cost of their return passage. These changes were influenced by the economic downturn after 1884 evidenced, for example, by an increase in the cost of production on estates. In this period also, there was an increase in the cost of immigrants' transportation to and from India. The landowners therefore increasingly clamoured for time-expired Indians to remain in the region and form a permanent labour force instead of opting for repatriation. Permanent settlement, they believed, would compensate for the high cost of importing indentured labourers.

The result of the change in the planters' attitude towards the importation of female immigrants was that by the late nineteenth and early twentieth century more women were being imported. On the *Chetah* in 1880, there were 112 females (nearly 44 per cent of the total) and 256 males. Females comprised 31 per cent of the total number shipped on the *Hereford* in 1885 and 30 per cent on the *Volga* in 1893–4. Of the 2,130 imported on the *Moy*, *Erne* and *Belgravia* in 1891, females totalled 689 or 32 per cent. On the *Belgravia* which imported 1,050 in all, females numbered 360 compared with 690 males (Shepherd 1994: 50–2; Shepherd 1995: 237–8). On some ships in the nineteenth century, the proportion of women landed in Jamaica even exceeded the stipulated female–male ratio of 40:100. For example, in the 1876 shipment the female–male ratio was 46:100 and it was 43:100 in 1877/8. Between 1905 and 1916, the percentage of women on each ship which arrived ranged from 22 to 30 (Shepherd 1995: 238).

Recruiters were not only encouraged to obtain more women, but more women 'of a respectable class', preferably as part of families. This meant excluding single, unaccompanied women. This was because there had developed an erroneous notion in India that single women were forced into prostitution in the colonies. Some visitors to Jamaica even seemed to have shared this belief – however unfounded. One H. Roberts, a noted opponent of immigration, claimed in 1847 that 'the utter disproportion of females in each locality tends greatly to the increase of vice and immorality' (C.O.R. 1847–1917a: 318/173). Lal and McNeil later agreed on this view of the existence of prostitution, though they disagreed that it was widespread. According to their report, 'of the unmarried women, a few live as prostitutes whether nominally under the protection of a man or not. The majority remain with the man with whom they form an irregular union.' They attributed this to the fact that some women were 'constantly tempted into "abnormal" sexual behaviour by single men with money. But they (the women) are open to temptation as on all estates there are single men who have more money than they need to spend on themselves alone' (see JT, 8 May 1915).

The Acting Protector of Immigrants in Jamaica also claimed that prostitution was noted among some Indian women in the island; so while agreeing that more women should be recruited, he warned that these should be of a 'better class'. According to him, 'it is no use increasing the proportion of women if they are to be picked up off the streets. They will only lead to further trouble as these women go from man to man and are ceaseless cause of jealousy and quarrels' (C.O.R. 1847–1917a: 571/1). It was in an attempt to induce women of a 'better class' to emigrate that landholders tried to reduce the period of indenture for women to three years. They believed that a shorter indenture and the promise of domestic life thereafter would be attractive inducement for the women and for their husbands. But not only were indentured men unable to afford the cost of maintaining their wives on account of the low wages they received, but Indian women demonstrated a preference for wage labour over uncompensated labour in the home.

Family emigration was supposed to help to solve the 'problem' of the emigration of too many single women. Before the early twentieth century, family emigration had been discouraged on the ground that this necessitated the importation of a large number of children who would increase the risk of epidemics, raise the mortality rate and delay the sailing of ships. This last matter was a perennial cause of concern, judging from the correspondence of the Protector of Emigrants in Calcutta in which he often produced figures to show how much delay could be caused by any unusual illnesses.

The fact that only children aged sixteen years could be indentured and that in many cases women would not emigrate without their children, had also been a deterrent. Towards the end of the indenture period, the view was that 'the emigration of whole families will be encouraged' (C.O. R. 1847–1917a: 571/6). While by the early twentieth century children of all ages were allowed to emigrate with their families, a preference was expressed for the recruitment of girls who would eventually increase the number of potential wives. Still, the numbers of boys and male infants shipped continued to exceed the numbers of girls and female infants. This was revealed in the sample survey of ships arriving in Jamaica between 1845 and 1916 which showed 508 boys and 333 girls being imported.

There was, predictably, some opposition to the emigration of families, on financial grounds. G. Grindle of the Colonial Office, in response to the recommendation of the Indian government officials, Chimman Lal and James McNeil, stated that 'the encouragement of the emigration of whole families, which in itself is a desirable feature of the scheme, will make the proportion of passages to working emigrants higher than under the existing system, especially as women will be under no obligation to work (C.O.R. 1847–1917a: 571/6). Despite the wishes of the planters, the majority of Indian women emigrated, not as part of a family (to conform to the mid-nineteenth century Victorian ideology of the 'proper gender order'), but as single women who had signed their own contracts for labour on the plantations. On the ship *Indus* which arrived in the island in 1905, for example, only 29 per cent of the women were noted as married and accompanied by spouses; 71 per cent were recorded as single or unattached. On the *Indus* of 1906, 33 per cent of the women were married and 66 per cent were single or unattached. Single Indian females continued to outnumber the married in the post-indentureship period. In the census of 1891, 2,851 out of 4,467 Indian women were recorded as single. Similarly the 1911 census showed 4,467 single Indian women, 2,479 married and 454 widowed. Forty-three women did not state their marital status. It should be stressed though, that there was some under-reporting of the marital status of Indians in the Caribbean in the censuses. Marriages performed according to Hindu or Muslim rites and which were not registered according to the legal requirements of the region, were not recorded.

Planters were also concerned about the tensions which developed among Indian men over scarce Indian women, and the violence against Indian women which resulted; for Indian men did not at first respond to the scarcity of female Indian partners by cohabiting with African-Jamaican women. There were frequent reports from men that Indian women were displaying a great degree of sexual freedom and independence. Some single Indian women reportedly changed partners frequently and seemed unwilling to marry any of the men with whom they developed sexual relations. This behaviour resulted in uxoricide and wounding of Indian women by jealous Indian men. Chimman Lal and James McNeil expressed the view that 'perhaps the best guarantee against infidelity to regular or irregular unions is the birth of children'. But the birth rate among indentured women remained low for the entire period of indentureship (JAR 1920).

In 1913 the Acting Protector of Immigrants in Jamaica supported an increase in the numbers of females shipped to the island as a remedy for the growing incidence of abusive behaviour towards Indian women. In a letter to the Colonial Secretary he reiterated that 'increasing the proportion of women would most likely reduce the number of cases of wounding and murder on account of jealousy, and be an excellent arrangement from the male immigrants' point of view as there would not be such a dearth of East Indian women as there is now on a good many estates' (C.O.R. 1847–1917a: 571/1).

Despite the attempts to increase the numbers of women in the island, the female Indian population in Jamaica was outnumbered by the male Indian population for the period of indentureship, as is indicated by Appendix 1 which is based on the population censuses. It is clear that up to the end of indentureship in 1921, Indian women were still less than 50 per cent of the total Indian population, though the proportion had improved from 31.5 per cent in 1871 to 45.2 per cent in 1921.

Location of Female Indian Plantation Workers

Most Indian women resided in rural Jamaica, reflecting the pattern of location of agricultural units on which they were indentured. The 1861 census detailed the distribution of the population by parish; but did not carry details on ethnic distribution as opposed to parish distribution by 'colour'. However, the details on the 'native country' of the inhabitants enabled an analysis of the distribution of the population from India by parish. It showed the majority living in Westmorland, Metcalfe, Clarendon, St. Mary and Vere (Higman 1980: 31–9). In 1871, the census showed a slightly altered parish distribution. The total Indian population in that year was 7,793, comprising 5,339 males and 2,454 females. The majority of Indian women were settled in Clarendon, Westmorland, St. Thomas-in-the-East, St. Catherine, St. Mary, St. James, Hanover, St. Elizabeth and Portland.

In 1881, the Indian population totalled 11,016, of whom 4,075 were female. The parish distribution in order of numerical importance was: St. Catherine, Westmorland, Clarendon, St. Mary, St. Thomas and Portland. Similarly, in the 1891 census, Indian women numbered 4,467 or 1.6 per cent of the total population of Jamaica. The majority were to be found in Westmorland, Clarendon, St. Catherine, St. Mary, St. Thomas, Kingston and Portland. From 1911 to 1943, St. Mary was the leader in terms of numbers of Indian females. One reason for the changes noted in the parish distribution pattern was that after 1891 the banana-producing instead of the sugar-growing parishes employed the majority of Indian labourers. On the whole, though, the censuses revealed a distribution which reflected a preponderance of Indian women in the rural areas, whether in the sugar- or banana-producing parishes. This rural bias was most marked up to 1881, and more notable among Indian women than among other female ethnic minorities. The census of 1881, for example, revealed that Chinese and white women – few of whom were involved in agriculture – were concentrated in Kingston and St. Andrew.

A shift took place only after 1943 when the rural to urban migration movement was reflected in the growing numbers of Indian women in the urban centre. This trend had actually started after 1881, a reflection of the trek of ex-indentured immigrants from rural parishes, but it intensified after the 1940s. The parishes experiencing the highest drain were St. Mary and Westmorland (see Appendix 2).

Women, Gender and Indentureship

Indian women began their experience in Jamaica as indentured labourers. One of the consistent features of colonial and imperial organization of migrant labour, the indenture or contract system, provided a means of retaining labour in the short and medium term and an institutional framework to facilitate the further movement of labour in the post-slavery period. At the inception of labour migration, contracts were for only one year, with renewal being optional. The period of contract for men and women was later extended to three years. By 1870, immigrants were given five-year contracts with repatriation due only after a further five years of continuous residence in the island.

Archival records yield more data on issues relating to fertility, the sexual disparity in migration schemes and male–female social relations than on the gender differences in the working condition of immigrants, specifically the extent to which they were subjected to sex-typing of jobs and gender discrimination in wages for equal work. But from the data available, it is clear that female Indians were subjected only to a limited form of the sex-typing of jobs according to which women were confined to service industries and men to agricultural field or factory positions. This sex-typing of jobs under capitalism was one of the forms of the sexual division of labour which European colonizers attempted to replicate in the Caribbean. It was traditionally created by the interaction in capitalistic society between the family and public economic life. But as Indian women could not be confined to the private sphere as wives of indentured men, and as there were insufficient openings for domestic servants in the scaled-down planter households of the post-slavery period, landholders were forced to use them in the fields.

Female Indian plantation workers were placed in one of three gangs on sugar and banana estates. The gangs were headed by males – usually African-Jamaican.

As during slavery, placement in gangs was determined by age, physical condition and gender. In the period of indentured labour migration, however, race/ethnicity was added to the other criteria for gang allocation; for invariably African-Jamaican workers were placed in the first gang (Shepherd 1994: 57). Up to 1880, Indian women were indentured on sugar estates. After 1880, they were primarily located on banana plantations. This shift was explained by the post-slavery decline in the sugar industry and the concomitant expansion in the banana sector, some sugar estates being turned into banana plantations. From 113 banana plantations in 1893, for example, the island had 435 by 1910. By the following year, 47 per cent of Indians were located on banana plantations in contrast to 39 per cent on sugar estates and 2 per cent on livestock farms (Shepherd 1994: 120).

The proprietors maintained a gender division of labour in non-field occupations. Thus while Indian women were confined to field labour and domestic service much as enslaved women had been, they were not given the factory jobs or the skilled artisan positions which were deemed suitable only for men. The few surviving plantation records indicate that indentured women had a narrower range of tasks on the sugar estates and banana plantations; and they were subject to discrimination in wages. They came to Jamaica during the operation of a system where men began to be paid more than women in spite of the experience during slavery that women survived the plantation experience better than their male counterparts. The contracts signed in the nineteenth century indicate that women were paid 9d.[pence] for a nine hour day and men 1s.[shilling] for the same number of hours, though not always for the same types of tasks (C.G.F. 1845–1916: 1B/9/3). But the acceptance of a differential rate of pay seemed to have been part of the patriarchal thinking of the period; for the wage differential was made an integral part of the indenture contract even before any tasks were allocated. In any event, the existence of a wage differential was predicated on the assumption that women's work was not as valuable as men's. In 1909, the Protector of Immigrants, Charles Doorly, informed Governor Sydney Olivier that 'during the first three months of their residence in Jamaica, immigrants are paid a daily wage of men 1/- and women 9d. (a day of 9 hours); 2/6d per week deducted for rations in the first three months' (C.G.F. 1845–1916: 4/60/10A/29, 1909). At the end of three months, theoretically, immigrants could ask to go on task work at rates of pay approved by the Protector; but in any event, it was stipulated that the rates for task work should allow immigrants to earn at least the minimum rates of 1s. for men and 9d. for women. In many cases, the tasks given to female workers were less remunerative than tasks given to men. Weeding, a low-paying task, was traditionally given to women, and many Indian plantation labourers were put in the weeding gang. The only exception was 'heading bananas', which paid 4–5s. per 100 bunches to both men and women. It is not clear from the sources whether men carried fewer or more bunches of bananas on their heads from the fields to the railway siding or the wharf.

On banana plantations, the most remunerative tasks, apart from 'heading bananas', recorded in work allocation books were: forking, trenching, ploughing, lining, circling and cutting. 'Trenching' paid 2–3s. per day and 'forking' 2s. an acre; but not all of these tasks were made available to women. Some men could earn up to 10s. per week from some of these tasks. Picking cocoa, a job that females did, paid 2d. for every 100 pods picked (Sanderson Commission: 1910). On sugar estates, as long as African-Jamaicans were available, they were given the more remunerative tasks. Less remunerative tasks were given to Indian men and the least remunerative to Indian women.

Even when task work was chosen over day labour, female immigrants failed to increase their wages significantly. On some properties, women even earned less than the stipulated rate. Lal's and McNeil's report contains quantitative data which will serve to illustrate this point. These data show that wages received by Indian plantation women in St. Mary ranged from 4¾d. to 9½d., though the upper level was hardly observed. On the thirty-six St. Mary estates in the survey, only two paid the women between 9d. and 9½d. per day or per task. These wage rates represented roughly one-half to two-thirds of the wages of indentured men (Lal and McNeil 1915; Shepherd 1994: 141).

A report on wage rates in 1919 showed that women earned an average of 6s. 11¼d. per week while men earned an average of 9s. 10½d. In 1920, men earned an average of 12s. and women 8s. 6d. per week. The Protector of Immigrants, from time to time, identified outstanding immigrants who earned above this average. Three women – Dulri, Inderi and Jaipali – all earned above 12s. per week in 1920; but in every case, the wages of the outstanding male workers identified exceeded those of the outstanding women – between 16s. and 18s. per week (see JAR 1920). In addition to earning lower wages, the records show that, but do not explain why, female workers were faced with a higher level of expenses than their male counterparts. At a conservative estimate based on rough statistics supplied by Chimman Lal and James McNeil, it would seem that the annual expenditure for females was 76 per cent of annual wages compared with 57 per cent for males (Shepherd 1993: 246).

Women with young children experienced further problems which affected the number of hours they worked and the wages they received. Where neither nurses nor creches were provided, indentured women often had to carry their infant children to the fields. This handicapped them in their jobs and could affect their productivity and therefore the amount of wages they earned. This was the complaint of women on some of the estates visited by the Acting Protector of Immigrants in 1913. He stated that 'recently when I visited a certain estate the indentured women complained to me that it was impossible for them to do a good day's work if they had to take their children to the fields and look after them there' (C.O.R. 1847–1917b: 571/1). A nurse was employed to look after the children and relieve the mothers of childcare responsibilities during working hours after the Protector appealed to the manager on behalf of the women.

On another estate where similar complaints were made by the female workers, the manager agreed to build a creche and employ a nurse to care for the children of immigrant women while they worked. The Acting Protector expressed his wish that, 'all employers of a large number of indentured immigrants ought to be willing to do something of the kind as a great deal more of the time of the women who have children would be available for work' (ibid). But not many estates adopted this practice, arguing that it was too much of an added expense for proprietors.

The result by the end of indenture was a general picture of 'persistent poverty' painted by those who visited the island and observed the conditions of Indian plantation workers. Such conditions were, arguably, initially worse than those of Indians elsewhere in the Caribbean. For example, J.D. Tyson (1939), an officer on deputation with the Moyne Commission which investigated conditions in the region after the 1930s riots observed that,

> the Indians in Jamaica struck me as the most backward, depressed and helpless of the Indian communities I saw in the West Indies. Their poverty is illustrated by the almost complete absence of property of any kind in their barracks and has the further unfortunate effect that their children are undernourished and are kept away from school for lack of suitable clothing to wear there.

Tyson was particularly concerned about the status of Indian women who, compared to their male counterparts, 'seemed to go for weeks without a single day's work', and whose welfare seemed not to have attracted much attention.

Women, Gender and the Debate over the Continuation of Indentureship

It was only when the continuation of the system of indentureship seemed threatened by rising costs of importation and repatriation in the early twentieth century that some improvements in the conditions of female indentured servants were suggested. Concerns over a possible labour shortage if labour migration ceased also caused the early twentieth-century immigration rhetoric to reflect a greater pro-natalist stance. Thus, just as the situation of enslaved women featured prominently in the emancipatory rhetoric of the 1820s and 1830s, and just as the improvement of their conditions was enshrined in the amelioration proposals to stem the tide of anti-slavery resistance as well as improve their fertility rate, so gender considerations were critical in the debate over the system of Indian labour which was to replace indentureship.

The discussions over the system of labour to replace Indian indentureship surfaced in the years after the First World War. It was suggested that the emigrants' agreement should be in the form of a civil contract rather than an indenture contract and that the term of contract should be reduced to three years. But the conditions of servitude for females were put at the centre of the debate. Suggestions were now made for women labourers with three children under five years to be exempted from work, subject to the approval of the Protector of Immigrants. It was also proposed that:

> any woman labourer may receive an exemption from work for any particular period either by agreement between the employer and the woman and subject to the approval of the Protector of Immigrants or on the Certificate of the Immigration Department. During advanced pregnancy and after childbirth, a woman may be exempted from work for a period not exceeding six months. Immediate steps should be taken to require the issue of free rations to pregnant and nursing women for a period not less than six months. (C.O.R. 1847–1917: 576/1,1916)

The inducements to be held out to male labourers, though, were linked to efforts to improve their economic welfare. It was suggested that any new scheme of Indian labour after the First World War should include provisions to make land available to male labourers. The recommendation was that all possible steps should be taken to require employers to provide small garden plots of one-tenth of an acre of land for each male labourer and facilities for labourers keeping cows. A larger acreage – one-third of an acre – should be given to those male labourers who were more industrious. This land was to be given after the first six months of labour in the island (ibid). No such considerations were given to Indian females, who in fact were to be encouraged to focus more on family – their 'proper role in life' – rather than on work outside of the home.

The Conditions of Post-migrant Indian Women

Choice, family ties, the severing of links with India after emigration and tardy repatriation arrangements, all combined to cause the majority of Indians to remain in Jamaica as settlers, the majority continuing as plantation workers. This was despite the fact that on the expiration of their contract, Indians were free to move out of low-paying plantation labour and seek more remunerative occupations. After ten years in the island, they could also access free or assisted passages to India (once they had not accepted the commutation bounty), or (despite the obstacles) seek higher-paying jobs in Cuba and Central America as thousands of African-Jamaicans had been doing since the 1880s. In reality, illiteracy and lack of access to land kept them tied to rural plantations as agricultural labourers. Only a minority were able to become subsistence farmers on their own plots of land. Some Indian men and women had been helped by the receipt of plots of 10 acres of marginal land in lieu of repatriation; but cash and land commutation grants had been discontinued by 1914. In the grants made in 1903/4, forty-eight women had received land, 67 per cent of them being single women. In the 1906 allocation, sixty-nine men and thirty-eight women got land grants (Shepherd 1994: 121–3). The women who came to the island after 1910 would have had to buy land on their own account or cultivate family land; and this was not only expensive but access was tightly controlled by the planter-class. Continued labour on rural agricultural estates was the only option for many Indian men and women, only a minority making it in the non-agricultural arena up to 1943.

In 1891 when the Indian female population numbered 4,467, 58.4 per cent were agricultural labourers. A further 399 or 9 per cent were general labourers, 151 were shopkeepers and 113 were doing household/domestic jobs. Only a few were numbered among the professional, industrial and commercial sectors which in any case, were male-dominated. For example, there was only one Indian female teacher. Appendix 3 illustrates the nineteenth-century occupational pattern.

According to the 1911 as well as previous censuses, the majority of Indian females worked as agricultural labourers mostly in the banana and sugar industries (see Appendix 4). On the all-island level, there were 49,116 females in agricultural labour, including 3,734 out of 7,452 Indian women. A significant proportion worked as domestic servants. On the all-island level, there were 35,701 domestic servants of which women made up 30,316. Of 188 Indian domestic servants, 134 were women; an additional ten Indian women worked in other domestic or personal service.

In the 1921 census, 3,828 Indian women were engaged in agriculture, again, the majority on plantations. Only one Indian woman was represented among the 144 rangers and supervisors on agricultural properties in the island. Among the peasant farmers, male Indians dominated. Of 188 Indian banana farmers returned in the 1921 census, there were only thirty-one females. There were thirteen females out of sixty-two Indian cane farmers; two out of thirteen cocoa farmers; 107 out of 399 provision farmers; forty out of 110 rice farmers, and seven out of thirty tobacco farmers (even though more females than males worked on tobacco farms).

Employment opportunities for women increased only marginally as a result of male emigration from the 1880s. The migration wave was dominated by African-Jamaican men and the gap created by their emigration was increasingly filled by Indian men, with African-Jamaican and Indian women getting work where male labour was not available. Even so, any such new employment opportunities were available mostly in agriculture.

By 1943, as revealed by the official records (1943 census) 56 per cent of Indian women were agricultural labourers, most still confined within the latifundial environment of the plantation. In contrast, only 28 per cent of African-Jamaican women, 1 per cent Chinese and 1 per cent Syrian women were returned as agricultural labourers on plantations. By contrast, whereas 56 per cent of Syrian women and 49 per cent of Chinese women were in the retail trade, only 12 per cent of Indian women were similarly engaged.

When one considers the general absence of schooling among Indians, particularly among the girls, the occupational pattern becomes more understandable; for there seemed to have been a close correlation between the educational level of the wage-earning population and the industry to which they were attached. Of 2,145 school age Indian children in 1891, only 110 girls and 126 boys were attending school (see JCR 1871–1943). A combination of social and economic factors militated against a more significant school attendance among both African-Jamaican and Indian-Jamaican children. By 1943, 49 per cent of the Indian population was returned in the census as illiterate. This was in contrast to 28 per cent African-Jamaicans, 14 per cent 'coloured', 3 per cent whites, 14 per cent Chinese and 6 per cent Syrian (see JCR 1871–1943; Shepherd 1993).

The economic crisis of the inter-war and post-war years also had an impact on the conditions of agricultural workers. The world-wide economic depression, the cessation of emigration as an outlet for those seeking higher wages and better living conditions, the return of migrants to swell the ranks of the unemployed, the termination of government-sponsored repatriation of Indians, failure of the planter-class to heed the workers' call for higher wages and better local working conditions had all culminated in widespread unrest in Jamaica and the wider Caribbean in the 1930s (Post 1978; Shepherd 1994: 142). When J.D. Tyson visited the island in 1939, he reported that there was a scarcity of work among women who formerly found jobs on coconut, banana and sugar estates. Many women went for days without work at a time when close to 60 per cent of them depended on the plantation for employment. According to Tyson (1939: 6), 'the complaint of short work for Indians – all the time on the banana estates and out of "crop" season on the sugar estates – was general wherever I went. Indians on banana and coconut plantations were thus living in a state bordering on destitution. Those who got 2–3 days' work per week and made thereby from 3s. to 5s. were considered fortunate.' Indian women were discriminated against in the allocation of such scarce jobs. Tyson (1939: 33) explained why this was so. He admitted that Indians suffered along with the rest of the labouring population of the island from the general wave of unemployment; but that, in the case of the Indians, this had been made worse by 'a growing competition in fields hitherto regarded as his own, from West Indian labour returning from Cuba and elsewhere with some training and experience in estate work'. In addition, there were allegations that headmen on the estates who were generally African-Jamaican, tended to 'favour their own people in the distribution of piece work, especially where there was not enough to go around'. He added that 'unemployment is general in the island but Indian labour has been especially hard hit by limitation of production in the two crops in which this labour is principally utilized – sugar (owing to the quota) and bananas for which, owing to the prevalence of various banana diseases, coconuts have been substituted'. The substitution of coconut for banana cultivation reduced the demand for labour and deprived the labourers of the customary use of marginal estate lands for their own cultivation of subsistence crops. Indian labour was, therefore, displaced and lacked the fluidity to seek alternative jobs in non-agricultural fields. Some found outlets nevertheless in rural-urban migration. Predictably, the parishes experiencing the highest drain were the depressed agricultural parishes of St. Thomas, Portland and St. Mary – also the most contiguous to Kingston and St. Andrew. In 1943 when Indians numbered 21,393, representing 2.1 per cent of the total Jamaican population, 8 per cent resided in Kingston and 14 per cent in St. Andrew. They were still very much a rurally settled population, with 76 per cent depending on wage labour; and they were only 2.6 per cent of the total urban population. The majority of those living in Kingston and St. Andrew were female who sought economic opportunities in the cultivation and door-to-door sale of flowers, fruits and vegetables, or worked as domestic servants in the households of the urban elite. Indian women were 2,297 or 52 per cent of all Indians settled in St. Andrew in 1943. By contrast, 33 per cent of Chinese women lived in Kingston and Port Royal and 18 per cent in St. Andrew (Shepherd 1986 and 1994: 135).

Conclusion

In conclusion, I wish to admit that the task of uncovering the historical experiences of Indian plantation women is not an easy one; for the colonialist historiography has often been 'gender neutral', keeping the working class, the subaltern, 'mute'. Indian diaspora women's experiences have been constructed less by themselves and more by those who claim to speak for them. The sources are predominantly official; and the 1000 letters plus, from and about Indians, which are stored as part of the Protector of Immigrants' files in the Jamaica Archives, though representing a potentially rich source for listening to the voice of Indian women, were mostly written by others on behalf of the immigrants and settlers. Furthermore, from the perspective of this topic, of the 1,858 letters selected for analysis, 77% were written by, to, or on behalf of men; 14% concerned female immigrants and settlers specifically. The concerns of Indians living in Kingston and St Andrew predominated, accounting for 67% of the letters compared to 25% relating to Indians in rural Jamaica, not all relating to plantation labour.

From the records available though, it would seem that the Indian female experience of migration and indentureship was conditioned by the colonisers' own perceptions and prejudices, and by race and class factors; but it is undeniable that gender discrimination was a visible element. The landholders' preference for male labourers, and their irrational belief in the inefficiency of female labourers, led to a gender-specific immigration policy; and the sex-typing of jobs and the accompanying wage differentials all helped to shape the Indian female experience in Jamaica. The types of occupations of female Indians in the period after the end of the system of indentureship reflected the lack of an educational standard necessary to equip them for higher-status jobs. It also reflected the ideology of the day which sought to confine women to 'female type jobs'. Up to the 1940s, the majority of Indian women were still tied to rural estates as plantation labourers, receiving lower wages than their male counterparts. This weak position of women in the labour market encouraged by wage differentials and the promotion of the male breadwinner ideology – the same notion of the 'proper gender order' which missionaries tried so hard to instill into African-Jamaican women in the immediate post-slavery period – was still in place up to 1943. After years of working as plantation labourers, Indian women were less able than other immigrant women to experience upward social mobility. Yet, despite their socio-economic marginalisation, after 1943, some did manage to break free of the plantation system, using educational opportunities and commercial enterprises to seek economic autonomy. As plantation workers they also used strategies of resistance to subvert the attempts of the landholding class to subject them to the exploitative elements of the indentureship system.

Appendix 1.   The Indian Population in Jamaica: Male/Female 1871–1921

Census Year	Male	Female	Total	Per Cent of Females
1871	5,339	2,454	7,793	31.5
1881	6,941	4,075	11,016	37.0
1891	6,338	4,467	10,805	41.3
1911	9,928	7,452	17,380	43.0
1921	10,203	8,407	18,610	45.2

Source: Jamaica Census, 1871–1921.

Appendix 2. Distribution of Indian Females by Parish: Selected Years

Parish	1881	1911	1921	1960
Kingston & Port Royal	54	196	357	446
St. Andrew	196	219	292	3,018
St. Thomas	206	692	626	624
Portland	189	610	627	630
St. Catherine	824	1,145	1,439	2,056
St. Mary	659	1,865	1,856	1,652
St. Ann	21	11	25	68
Clarendon	732	1,070	1,215	2,313
Manchester	34	36	30	88
St. Elizabeth	139	77	102	158
Westmorland	779	1,275	1,461	2,649
Hanover	111	108	171	126
St. James	59	79	147	163
Trelawny	72	69	59	35
TOTAL	4,075	7,452	8,467	14,026

Source: Jamaica Census, 1881–1921.

Appendix 3. Occupations of Indians in Jamaica: Male/Female, 1891

	Men	Women	Total
Peasant proprietors	59	3	62
Overseers	15	0	15
Agricultural labourers	3,707	2,534	6,241
Attending stock on pasture	10	0	10
General labourers	523	399	922
Attending agricultural machinery or boilers	1	0	1
Merchants/Agents/Dealers	10	0	10
Shopkeepers	246	151	415
Shop/Sales/Clerk	39	46	85
Market Gardeners	60	50	110
Indoor Domestic Servant	35	50	74
Washers	0	26	26
Interpreters/Messengers	4	0	4

Source: Jamaica Census, 1891.

Appendix 4. Summary of Occupations of Indians in Jamaica: Male/Female, 1911

	Male	Female	Main Categories
Professional	35	11	Students, Nurses, Teachers
Domestics	91	152	Indoor house servants
Commercial	386	204	Barkeepers, Peddlers Shopkeepers, Store Servers
Agricultural	6,649	3,734	Wage Earners on Plantations
Industrial	165	78	Skilled Trades, e.g. milliners, washers
Indefinite and unproductive*	2,602	3,273	

Source: Jamaica Census, 1911.
* Included women working at home!

Appendix 5. Percentage Distribution of Indian, Black, Coloured and Chinese Men by Parish, 1943

Parish	Indian	Coloured	Chinese	Black
1 Kingston and Port Royal	7.7	18.1	32.9	7.4
2 St. Andrew	13.8	14.7	17.8	9.3
3 St. Ann	0.4	6.7	3.3	8.2
4 St. Catherine	13.8	5.8	6.9	10.5
5 St. Elizabeth	1.2	11.6	2.5	7.5
6 St. Mary	17.6	5.2	5.0	7.4
7 St. Thomas	5.8	1.9	3.7	5.4
8 Clarendon	13.2	7.2	6.9	10.4
9 Hanover	1.9	4.7	1.7	4.1
10 Manchester	0.7	6.8	4.3	7.9
11 Portland	5.6	3.0	5.1	5.2
12 Trelawny	0.7	3.3	2.4	4.0
13 Westmorland	16.4	7.1	3.0	7.1
14 St. James	1.2	3.9	4.5	5.6
Total Per Cent	100.0	100.0	100.0	100.0

Source: Population Census of Jamaica, 1943.

7

Gender Relations and the Plantation System in Assam, India*

Shobhita Jain

Introduction

Gender relations among the workers range from examples of extreme oppression of women to possibilities of limited degrees of gender equity. Without undermining the relevance of studies which bring out the structural dominance of men, I have examined the workers' everyday struggle which is both stark and vibrant and manifests their effective strategies for a sociality marked by horizontal gender relationships.

By gender relations I mean the whole range of social distinctions made between the sexes; either extending through the overall sphere of social life or confined to certain aspects. Both cooperation and antagonism characterize the distinctions of gender. Most gender differences are cultural projections onto biology; however, this does not mean that women's position can be explained in terms of biological make-up. Such explanations have already been contradicted by the evidence of the large variety of female behaviour. I look into such institutional arrangements as reflect social distinctions between the sexes within the tea garden[1] community. I have not felt obliged to use the terminology of 'domination by men' and 'the oppression of women', although such an interpretation may be warranted in many situations. Gender relations are here analysed in terms of an alternative ideology that women have about society and their place in it. I have argued that women also can be the 'subject' rather than being the 'other'. While deconstructing the feminist paradigm and its notion of universal subordination of women, I have articulated the historical and cultural contexts of the Assam tea gardens to highlight a specific and localized pattern of social relations among the workers.

* The author acknowledges support of the Indian Council of Social Science Research and the E. Horniman Anthropological Scholarship Fund, RAI, London, which facilitated research and field work carried out in Assam, India. Comments by the organizers of and participants in the Second World Plantation Conference at Shreveport, Louisiana, in October 1986, were also useful in writing this article.

Gender relations are discussed by me in the context of the family as an economic unit and women's as well as men's participation in household organization. It is useful to relate domestic labour to the total mode of economic activities of which it is necessarily a part. Women in the tea garden are not confined to the child-bearing and rearing aspects of life and therefore I do not treat them as agents of human reproduction alone; I take them primarily as manual workers, participating directly in productive activities of the tea industry and their domestic labour is marked by the delegation of responsibilities by senior women to junior members, mainly to unmarried daughters. Frequent role-crossing in the division of labour is another feature of household activities. Nimari[2] women have learnt over time to strike a balance, through reciprocal relations of an egalitarian kind, between their domestic and work spheres.

The term 'egalitarianism' here refers to relative equity in reciprocal relationships between the women and the men. Lack of rigidity in male and female roles in household organization and the high value of manual work for either sex provide spaces to women for a larger degree of visibility in decision-making in everyday life. As noted by Mintz (1974: 299) corporately employed manual workers 'stand in like relationship toward the source of their employment and in their interaction they evolve a pattern of reciprocal relationship with each other'. Nimari women's perceptions of themselves primarily as tea garden workers do not revolve around a view of themselves as having to do double duties, with one sphere of life being neglected on account of the other. They cannot conceive of a situation without employment. Of course, there are unemployed females in the labouring community, yet, the whole orientation of a female on the tea garden is directed towards work and her organizational abilities as well as her relations with other members of her household mean that she is prepared for regular employment. Men and women share rights and responsibilities in and outside the family contexts because they must, in order to survive and reproduce. The spheres of production and reproduction are seen here in a continuum.

It is not that there are no quarrels or points of dissent among them; for example, when both are drinking 'home brew'[3] fights often occur. But Taramoni, a plucker, said, 'Who knows who gave more thrashing to whom? When you are drunk you have no control over your actions. But in a normal situation no man would ever beat a woman for any reason and for that matter, no woman would beat a man either. We do not ever beat even our children.' One is aware that the relative equity in gender relations in Nimari is not a product of any conscious planning, as is perhaps being attempted by women and men in modern societies. My argument is that only the force of egalitarian values has helped them carry forward and effect their survival against all odds. In the following case-study I have shown that a severely limited measure of equity among the tea garden workers emerges as their strategy for survival in an intensely oppressive social environment. By strategizing a pattern of co-operation and mutual dependence between the sexes, at the level of their sub-culture, the workers have attempted to cope with and adjust to the authoritarian structures of plantation society (Jain 1988).

While discussing mechanisms of survival one is neither celebrating the plantation system nor denying or diluting the unfree nature of social relations of production. Brass and Bernstein (1992: 5) try to prove that the survival strategies' framework denies or dilutes the unfree nature of wage labour and hence it entails a 'revisionist characterization of the plantation workforce'. In the case-study presented here, the 'adaptive' approach of the survival strategies argument does not imply freedom of making choices. The very fact of women having to strategize their survival means an utter lack of free choices and thereby severe constraints and limitations to further growth of equality among the workers. Further, Lieten and Nieuwenhuys (1989: 8–13) have linked the notion of survival strategy to analyses of emancipatory development of individuals, whereas I have shown that the form of survival which occurs in the context of a tea garden life is not an individual matter. Long-term socio-economic survival is a product of the workers' collective strategies. It is, of course, related to physical survival. There are, however, hardly any patterns of activities among the workers that can be characterized as efforts to improve their lot. The tea garden workers in general, and women among them in particular, are more or less trapped in plantation enclaves and strive to survive as well as they can. The pattern of limited gender equity that obtains in the tea garden labouring community is inadequate to negate the exploitative relations across classes and status groups in the system. Furthermore, were the tea garden women to move outside the plantation enclave, there would be little scope of continuing even the limited degree of gender equity because, compared to their men, tea garden women have almost no possibilities of finding alternative jobs. Moreover, norms of male authority and domination in society in general are likely only to erode whatever little parity they experience within the confines of tea gardens.

Historical and Contemporary Context

Capitalistic growth of the world economy in the mid-nineteenth century demanded the increasing use of migrant labour which, for Assam tea planters, was cheaper to acquire than to employ local labour. Historical accounts of the colonial plantation regime in Assam bring out the dominance of planters in the state administration and their desire to get labour at the cheapest possible cost (see Guha 1977: 9–18; Jain 1983: 262–84; Siddique 1990: 191). Facilitated by the colonial administrative machinery, the European planters managed to recruit the poorer sections of village communities from different states adjoining Assam. Alienated from the means of production and separated from the product of their labour, they produced a commodity for the world market by carrying out routine agricultural operations. While analysing gendered power relations within the labouring community one cannot afford to set aside the fact that in Assam tea gardens, as in most plantation settings around the world, the industrial organization, that is, the institutional arrangement connected with production and the market, established a pattern of authority and control over almost all aspects of the lives of the people within its territorial limits. Consequently, the sub-culture of the resident labouring community shaped in a particular manner.

Although a community of heterogeneous origins, both female and male workers in the Assam tea gardens formed a common class in relation to the combined supervisory and managerial classes. The general manager and his staff (referred to hereafter as higher participants) stood in a position of super-ordination vis-à-vis the labourers who were and are at the bottom of the occupational and social hierarchy. In the entire history of Nimari, there has not been a single case of mobility from lower to upper classes. In the 'line of order' from manager to labourer the class and status positions are both rigid and closed. Considerations of home, locality, nationality and traditional values have been retained or ignored according to the demands of the economic organization of international tea trade. This is our broad frame of reference to discuss gender relations within the labouring communities of Assam tea plantations.

Owing to persistent conditions of economic depression in the 1930s, planters had shifted their burden on to the labourers with disastrous consequences to their day-to-day survival. Both female and male workers had subsequently joined hands with oil workers, creating a wave of strikes (Guha 1977: 242–6). Among them was the famous strike on 30 July 1940, at Allenpur Tea Garden in Cachar, where 200 women workers stopped work and demanded higher wages. Women have been reported taking decisions during crisis situations and leading labour processions not only in the past, but also in the post-colonial tea gardens of the Dooars, in North Bengal (see Bhowmik 1981: 129). Nevertheless, such features of the colonial plantations as use of extra-legal authority by managerial staff, and the hierarchical socio-economic power structure, continue to characterize the post-colonial plantations (see Jain 1982: 106–40). Our discussion of reciprocal relations among the workers, particularly those associated with gendered power relations, shows that the dominant norms of male domination do not find full play in labour 'lines'.[4] Consequently, the process of family formation has over time provided, in the moral sense, 'wholeness' or a sense of being complete in oneself to the tea garden women (cf. Breman and Daniel 1992: 285).

For the Nimari workers, originally belonging to pre-industrial societies, now working in an industrial set-up for their physical survival, the whole meaning of life has changed from one set of values to a completely different identity set and status. Their work is central in the sense that it provides them with wages to survive and also it is central as a source of status or identity in the wider society. The 'labour' identity of Nimari women reflects an absence of the institutionalized patterns of male authority.

The 'Labour' Identity of Nimari Women Workers

Regarding status within the labouring community, we find that basic or ascribed roles of Nimari women do not follow the conventional sex role-model, in which gender-roles, based on biology, are basic. Congruency between a woman's nurturing role and occupational role is just as great as in the case of a man in the Nimari working community. Organization of work in Nimari is based on social units specially recruited for that purpose. This is in sharp contrast to the work in small-scale tribal societies where work is performed by groups of people not usually brought together solely for that purpose. In that case the family or even a village may act as a unit in the production process. From the background of this kind of work, the migrants were required to shift to an entirely different concept of work under the colonial planters. Over time, the workers have displayed a remarkable capacity of adjustment and adaptation to new surroundings and requirements. The process of acculturation here is very complex not only because Nimari workers lead an enclosed, formally administered round of life and undergo a similar experience of 'total institution', but also because the heterogeneity of groups, originally recruited to Nimari, as on most tea gardens of Assam, is unparalleled. After a century of experience of plantation life, the 'coolie' (pre-Independence) or the 'labour' identity of tea workers in Assam has now reached a stage of consolidation both in the eyes of the workers themselves and others in Assam. The specialized style of life (see Jain 1991) and the social relations of production on the Assam tea plantations reflect a process of homogenization of the heterogeneous cultural patterns of the labourers, who comprise the second and third generations of migrants from rural and tribal areas of the states adjoining Assam, mainly Bihar, West Bengal and Orissa. Table 7.1 shows the distribution of tribal and caste groups by the number of households, with a preponderance of tribal households over those belonging to different castes. It also shows the multiplicity of geographical areas from which the migrants came.

As said earlier, cultural and other factors from the workers' past are retained and perpetuated on the tea garden to the extent that they aid the process of production and capital accumulation. The economic system, in this sense, dominates or encompasses the socio-cultural dynamics of the labouring community. One is not here positing crude Marxist determinism, rather one is linking the fact of the almost total non-existence of workers' access to resources to their survival strategies within the household.

Table 7.1.   Tribal and Caste Groups of Nimari by Areas of Origin

Area	Number of Households				Number of Named Groups from each area			
	Tribe	Caste	Total	Per Cent	Tribe	Caste	Total	Per Cent
Assam	5	2	7	1.31	2	2	4	7.84
Bihar	184	105	289	54.22	14	22	36	70.84
Orissa	76	2	78	14.44	5	1	6	11.76
Andhra Pradesh	—	18	18	3.37	—	1	1	1.98
West Bengal	—	27	27	5.06	—	3	3	5.88
Tanti (amorphous)	—	90	90	16.88	—	1	1	1.98
Unknown	—	—	24	4.50	—	—	—	—
Total	265	244	533	99.78	21	30	51	99.98

The following discussion of the main features of the Nimari tea garden workers' community delineates the nature of gender relations in the context of division of labour in 'on' and 'off' work situations. Nimari is one of the three production divisions of Behula Tea Estate in Sibsagar district of Assam. Ethnographic data here relate to conditions during the period 1978–81.

The Nimari Labouring Community

The Nimari division employs 860 workers on a permanent basis, in addition to a varying number of temporary hands.[5] Both permanent and temporary workers are rated as unskilled manual labour, and are paid by day or by piece rate. Exception to this norm are forty-one salaried workers, who are paid monthly, but at the same rates as other workers. All permanent workers live in labour 'lines' provided by the estate. Female workers, if not living with parents or spouses, are allotted houses in their own names and occupy them in their own right.[6]

On Nimari, workers are divided into three gangs of female, male and minor labourers. Anyone above the age of twelve can start work and there is no specific age of retirement. Among the wage-earning labourers, 48.1 per cent are female adults, 47.3 per cent are male adults, 2.8 per cent are female minors and 1.1 per cent are male minors. Together, they account for 95.5 per cent of the workforce, and to them are added the forty-one salaried workers who are all male. In terms of carrying more authority and power within the labouring community, among the salaried workers the occupations of sardar[7] (foreman) and *chowkidar* (watchman) are more significant than those of carpenter, driver and washerman. The Nimari management has, so far, not appointed a single woman to any of these positions, with one exception when, in the early 1930s, one tribal woman was appointed as a recruiting sardar and she brought at least three recruits to Nimari from her village in South Bihar.

Opportunities for occupational mobility within the estate are almost non-existent for the workers. Currently, the management gives preference to persons of Assamese origin, in contrast to the previous practice of recruiting middle-class Bengali men, as supervisory or technical staff. The chances for paid work outside the gardens are scarce and the workers have to somehow devise strategies for survival within the economic constraints of the plantation.

The labourers work for eight hours per day for six days of the week. Their work begins at 7.00 a.m. in summer and at 8.00 a.m. in winter.[8] In 1981, the Nimari management paid its female and male workers at an equal rate of Rs. 6.64 for 8 hours of work.[9] In addition, the workers receive benefits[10] which are considered, by the employers, to compensate for the low level of cash wages.

In tea production most labour is required in the field branch, and only nominally, in the factory, office and medical branches. The field branch is exclusively concerned with tea cultivation and harvesting, while the factory takes care of its manufacture. Most adult and minor female workers engage in the task of plucking tea leaves, while most adult and minor male workers carry out field work, involving weeding, pruning, digging trenches, spraying chemicals over tea bushes and so on. Field work is considered to comprise heavy manual tasks while plucking is generally described as a skill that women have achieved through their nimble fingers. Since the task of plucking is the single most costly operation on the garden, and women perform this task, employing them is a matter of necessity to the planters. Clearly, despite rationalization, the tea industry is still heavily labour-intensive and in the task of plucking two leaves and a bud, scientific technology is not of much help. Rather it is the human skill provided by women that is crucial for producing high quality tea. It is true that women are not given any special benefits as a reward for their special contribution to the tea industry, yet their importance for the industry cannot be denied. As shown in the following account, they participate in a labour process that is characterized by somewhat low institutionalized entrenchment of male domination.

Labour Supervision

In both 'on' and 'off' work situations, women labourers have close contact with male supervisory staff, specially the sardar. Selected from among the labourers, foremen or sardar are appointed to supervise the work of female, male, minor and temporary workers. The real responsibility of maintaining an adequate level of productivity lies with the higher officials; they constantly instruct the sardar who is, in turn, responsible for bringing the labourers to the fields. Sharing the management's responsibilities, the sardar becomes the 'management's man' in the eyes of ordinary labourers. In the role of recruiter, a labourer, who returned home to bring his own village people to the garden, was known as sirdar or sardar. In vernacular Hindi, sardar means 'leader'.

All pluckers, without exception, said that their sardar did not scold them for being late so long as they reached the fields within half an hour of the reporting time. As long as a gang member was careful not to embarrass the sardar before his superiors and, for example, informed him in advance about taking a day off – thus allowing him a chance to make an alternative arrangement – the sardar did not exercise the punitive powers given to him by the estate manager to ensure work efficiency. In order to maintain the element of goodwill between himself and his gang members, upon which his job largely depended, he would let a daughter help her mother or a mother nurse her baby whenever there was the need. If a woman did not feel well enough to continue her work he would allow her back to the 'lines' without reporting the matter to the *babu* (member of the office staff).

Women are uniformly excluded from the group that has been created by selecting persons who can be trusted to carry out the manager's orders and maintain discipline among the workers. When asked to suggest reasons for this situation some of the women workers explained that, since they have the primary responsibility of nurturing the small children, they cannot be expected to exert disciplinary action on the labour force. Because of household chores they often arrive late at the workplace, and if they are themselves not punctual they cannot expect others to work under them. At the same time, as Padmaboti pointed out, the sardar has to be careful lest his gang members become antagonized and approach the manager complaining of his unreasonable or vindictive behaviour. Not only this, he has to keep them in good humour. In 'off' work situations the sardar felt a moral obligation to attend to the several needs of workers of his gang. If there was someone ill in the family of a worker he made it a point to visit her 'line' and ask about the person. On such occasions as a wedding and other life-cycle rituals he offered help. Jarkoo Sardar, for example, provided the patromax light for the wedding of the daughter of Sukurbati of his gang. He also went to ask whether anything else was needed. Some sardar are also known to lend money to their gang members, though this activity is more closely associated with bonds of kinship and neighbourhood than with gang relationships.

However, the 'lines' do not now have clear-cut patterns of authority vested in the hands of a few. Some sardar are vehemently criticized as being the management's henchmen who have few scruples about their conduct. 'Such persons do not remain in favour for long and are soon succeeded by more unscrupulous lot', said Sukurbati, the daughter of an old sardar. The position of sardar is, therefore, not one of pure privilege, and only those who can command the goodwill of their gang members hope to continue for long. Such people do not consider themselves as being over and above the other workers.

It is obvious that on Assam tea gardens the sardar has not the same role as a *kangani* on Sri Lankan/Malaysian plantations (see p. 12). In the latter case, institutionalization of the *kangani*'s position of authority in the work situation becomes combined with his position of male authority in the family. Whereas on the tea gardens levelling down of the earnings in various categories of workers pushes them towards proletarianism and towards relations of parity among themselves. Clipping of the work-related authority vested in sardar (males) seems to have taken away the edge of superiority of male positions in the community.

Pooling of Labour and Access to Cash

The landless labouring class, which had migrated to the Assam tea gardens, had a tradition of pooling male and female labour to earn survival wages. Women in this class, as a normal practice, always combined the roles of mother and worker. On tea gardens, women and men continued to share the economic responsibility of providing sustenance to their families. Women's direct participation in production precludes the prospect of their confinement to the 'private domain'. In addition, direct access to cash wages gives them control over expenditure in the household and reduces their dependence on men for daily requirements of cash. No worker in the estate is allowed to collect another worker's wages on her or his behalf, which is another reason for the absence of male control of women. In contrast, Kurian (1982) mentions that women on Sri Lankan tea plantations had almost no access to the cash which they earned. Access to cash gives Nimari women the chance to spend it, on their own, in the weekly market for meeting their families' requirements.

In 1981, a field worker, invariably a male, earned an average of Rs. 7.06 per day, a factory worker, again invariably a male, got an average of Rs. 7.37 per day while a plucker, invariably a female, received an average of Rs. 8.94 per day. Thus, female workers, mainly employed as pluckers, had larger pay packets at the end of the week. Yet, in the perceptions of the women, over a period of one year their average earnings worked out to be more or less equal to that of men. This was because women usually worked a slightly lower average number of days per year than men. Sombari, a plucker, said that a more important concern for her and all others in the 'lines' was how to match the total income of the household with its total expenditure. Within the garden premises a weekly market is set up and women can be seen going straight to the market after receiving their weekly wages on pay-day (see Jain 1988: 50–58). This is another area where division of labour indicates absence of male domination. In theory both women and men spend their entire weekly earnings on meeting the expenses for food and the survival needs. In practice, women plan and execute the pattern of expenditure in the household.

Control Over Earnings and Expenditure

Regarding control of the family purse, there is no hard and fast rule. Depending on the size, stage and composition of the household, it may be in the hands of either the woman or the man or both. Also, all working children spend a fair share of their income themselves, contributing only about 10 to 15 per cent of their earnings to the common pool to be spent on food items. This contribution is, of course, in addition to the subsidized rations of cereals to which each worker on a tea garden is entitled.

Working children with permanent jobs often initiate the process of setting up their own households by asking the garden authorities to allocate a separate 'line' and at the same time they start looking around for life partners. This is also the stage in the developmental cycle of families when conflicts over many issues occur between the parents and the working children. Often quarrels relate to such household chores as firewood collection and chopping, and filling of water. During a crisis situation such as illness or disability when extra cash is required, working children are sometimes unwilling to contribute. There are hardly any saving mechanisms operating in the labour 'lines'. In most cases of illness and subsequent unemployment, the workers face serious difficulty in meeting their regular expenses. For additional expenses during life-cycle rituals they need to incur debts. Since their ability to raise credit is extremely limited, they are forced to simplify many of their social customs. When it comes to sheer survival, they resort to the most common method of securing small loans by asking the estate office for an 'advance' against their earnings. This ties them further to the plantation. Reflecting on this cycle of 'advance' and its repayment through one's labour, Budhu Sardar once described his life in the tea garden as the life of a beast who earns, eats and sleeps without a trace of the kind of life his ancestors led in Chotanagpur villages. Responding to the quiver in his voice, Rondhai, his daughter, reminded him that he should not forget that he is so good at making walking sticks from ordinary bamboo poles. She asked him, 'Can beasts produce such craftsmanship?' Amidst muddy tracks, mosquito infested labour 'lines', bare minimum life supporting means, Rondhai was able to hold the old man's spirits high and divert his mind away from harsh realities. This made her whole and through her also him. The old man simply wiped his eyes and gave her a faint smile.

The Spheres of Production and Reproduction in a Continuum

Engaged in both the micro and macro production systems, Nimari women and men not only reproduce the labour force but also create the ideological base for maintenance of the exploitative system by providing the socialization and day-to-day servicing of the family members. The process of family formation for building up the more stable workforce is marked by the interplay between the economic constraints and the socio-cultural features of the tribal and caste groups. Currently, the ability to reproduce the workforce from within and its survival are two factors most essential to the efficient running of the tea garden. It is therefore imperative that their superiors ensure that the women and men not only work but also survive and reproduce. For survival, the workers are given just enough cash wages and subsidized rations of cereals; for reproduction, just barely minimum living conditions in the labour 'lines'. Thus the community is left on its own to devise ways and means of continuing the social relations of production. Their wage-level, determined by the estate on the assumption that each adult earns an income, is such that the women are obliged to work as well as reproduce the workforce.

The nature and type of marital union among the workers of distinct tribal identities and lower castes reflect a considerable degree of decision-making by the women. A recognized sexual union between a woman and a man in the Nimari labour 'lines' confers an acknowledged social status on the off-spring who belong to their respective tribe/caste and also to the 'labour' class in the industrial sub-system. *Bagan Bat*, the tea garden lingo, has two terms to describe marriage in Nimari. The first term, *bandobast*, refers to a marriage arranged by parents or by the marriage partners themselves. This involves the traditional rituals, ceremonies and customs of bride-price and 'milk-money'. None of the tribal and caste groups on Nimari observes the practice of dowry, while gifts, in the form of cash and/or kind, for paying bride-price and 'milk-money' are common at pre-wedding ceremonies. Almost always among the tribal groups, the bride-to-be formally gave her acceptance or refusal to the proposal of marriage. Generally, a person's first marriage is of the *bandobast* type, though the incidence of the second type, *rajikhushi* is also common in first marriages. Clarke (1979: 73–89), reporting the situation in Jamaica, has shown how legal marriage and common-law union are sharply segregated conditions and stages in mating. On Assam tea gardens, the two are closely linked. A *rajikhushi* marriage refers to a common-law marital union, and is contracted on the basis of one's free will and pleasure. The term usually denotes a person's subsequent marriages but can also apply to a first marriage in which customary gift-exchanges and community feasts are deemed to have been postponed for a future date, which, in many cases owing to the lack of sufficient funds, never take place.

When a girl and a boy decide to marry and the parents or elders in the family do not consent, they simply disappear for three to four days in the jungle surrounding Nimari and later come back to the garden to live as a married couple in their own house, if one is allotted to either, or with the parents, relatives or friends of one of them. The allotment of the house depends on the occupational status of the girl or boy, and hence an appropriation of greater independence in decision-making by the girl or boy and corresponding weakening of the parental authority in the family. Nimari parents expect and groom their children to start earning as soon as they are twelve as the estate stops providing the subsidized rations of cereals for children above the age of fifteen. Everybody, whether a girl or a boy, in Nimari labour 'lines', is expected to provide for the rice she or he consumes. This basic fact of life becomes the guiding factor in the pattern of mating and reproducing in Nimari.

Relative Absence of Marked Male Authority in Gender Relations

Predominant tribal labour on Assam tea gardens and totally non-tribal labour on most South Indian and Sri Lankan plantations are likely to account for widely different types of gender relations in the two regions of the same country. Among non-tribal workers belonging to Hindu castes one is likely to find extreme forms of sexual inequalities and prevalence of patriarchal authority over women (Samarveera 1981; Kurian 1982). In contrast the tribal character of institutionalized forms of reproduction among the Assam tea garden labourers seems to have given the women space for decision-taking (see Jain 1986: 189). Further, the causes, frequency and consequences of incompatibility in marriage are causally linked to the considerable incidence of female-headed households in Nimari (Jain 1988: 91). Let us look at this situation in detail. As said earlier, the socio-cultural practices of workers of tribal as well as caste groups have also been affected by the politico-economic constraints of the plantation system. To that extent, impact on the tea garden life has been that of steamrollering the sharply distinctive features of tribal and/or caste identities of the people. This has further impelled Nimari women and men to underplay traditional forms of marriage and its attendant rituals and ceremonies.

The garden workers would like to contract marriages within their tribal/caste groups. But because of the particular historical circumstances of estate living, many marriages have had to disregard the rule of endogamy. Discussing the marriages outside one's endogamous group, the Nimari residents explain that in the past there were no easy means of communication, and often a person was forced to seek a spouse within the garden, from another caste or tribe. Also, because of frequent epidemics in the early days of the tea industry in Assam an unusual number of people died on the estates, and in such cases many widows and widowers re-married, often to partners outside their tribes/castes. In several instances such unions were suggested and arranged by the estate manager on the basis of practical consider-ations. For example Dhapalu who had lost his wife and Santimoni who had lost her husband were asked by the manager to live as husband and wife and to take care of each other's children. Thus the manager saw to it that his garden did not suffer loss of its potential workers. An abundance of such examples seems to have taken away the pressure of male domination on family formation in the tea gardens. In the new setting there were hardly any means of controlling the behaviour of those who deviated from traditional norms, and since the new controllers of social sanctions (the estate management) positively approved the *rajikhushi* form of marital union, which served the management's goal of building up the future workforce, the practice of endogamy lost its rigidity and inter-caste, inter-tribal and inter-ethnic marriages found social acceptance in the labour 'lines' as well as in the tea garden as a whole.

Again, in matters of residence after marriage, patrilocal residence, and thereby the severing of close contact with one's family of orientation, is hardly a norm as a large number of Nimari women have married within the garden and neither the woman nor the man had to go out. Where a woman has permanent employment and her spouse is only a casual worker he joins her for, as a permanent worker she has the entitlement to a 'line'. The practice of neolocal residence, formed either by the woman or the man, is regulated by occupational status rather than by the traditional practice of patrilocal residence. There were 493 married women in Nimari in 1979, and of this number 113 were married within the garden and lived close to their families of orientation. Depending on her occupational status, in circumstances of marital discord a woman had the option of continuing to live in her existing accommodation. In practice, it was often found that it was the man who made himself scarce allowing the woman's subsequent spouse to move in. It does not, however, follow – see the speculation made by Dube (1986: xxxviii–ix) – that a daughter is likely to acquire the house allotted in her mother's name. Only the proximity of residences of mother and daughter, and hence the change of emphasis from patrilocality to neolocal residence, is stressed here. My point is to bring out the tea garden variations in tradition available to women. Even a limited degree of such variations begins to destroy the rigidity of traditional sex-roles in which women are universally found to be invisible, voiceless and incomplete.

As further evidence of tea garden women's social spaces, which are precariously situated within the interstices of an exploitative economic system, we find that there is no child marriage on Nimari as both types of marital union (*bandobast* and *rajikhushi*) require economic arrangements that imply that both the woman and the man, or at least either one of them, is permanently working in the garden. On Nimari, according to an old resident of the labour 'lines' of the 1930s and 1940s, a girl usually got married and began to live with her husband between the ages of thirteen and sixteen, and it was common to see a seventeen year-old with a child. By the 1980s girls' age at marriage had certainly gone up. In Nimari out of 493 married women only nineteen women had married before the age of eighteen, while eighty girls above the age of eighteen were still unmarried. It is expected that a man should be senior to his wife in age at marriage. As a result the married man is never below eighteen, but is invariably older than this as very few male workers manage to find permanent jobs before the age of twenty or twenty-one. Normally, it is only after he becomes a permanent worker that a man plans to get married.

Being able to find permanent work at a relatively lower age than the men workers, women in tea gardens are in a position to see themselves as holding independent status. In most agro-industrial organizations women are employed as secondary workers and housing is provided only to male workers and their families (Breman and Daniel 1992: 287–8). In Nimari, a woman worker's entitlement to a 'line' enhances her marriage prospects and continued residence near her natal kin ensures her their support in managing household chores and in providing her an active role in the decision-making (see Jain 1982: 419–22, 438–51 and 608–12).

Moreover, the idea of indissolubility of a marriage does not exist in Nimari. This does not, however, imply that most marriages are dissolved. As a norm, the majority of marital unions in the 'lines' have remained stable, and the taking on of a second partner by either sex has arisen only upon loss of a partner by separation, divorce, desertion or death. Several instances of married women eloping with other men, single or married, cases of temporary or permanent separation and of formal divorce show that the traditional power structure of patriarchy does not fully operate in Nimari labour 'lines'(see Jain 1982: 446–69). Soon after marriage, if a woman finds it difficult to adjust to her new life, she decides to break the marriage by returning to her parents. If a bride-price was paid for her it is normally returned to the man through regular deductions from her earnings. In all such cases, there is the formal recognition of break-up of the marriage, and this can be equated with the term 'divorce'. When bride-price is not returned, the man tries to get his wife's sister to live with him, or to get the money or goods back by appealing to the 'elders' in the community. It is said that though there may be a delay in repayment, there is never a forfeiture.

If a woman leaves her husband and children for another man, the community criticizes her for lacking affection (*mamta*) for them. To separate from one man to live with or to marry another, however, is not criticized per se. It is usually the woman who decides to break a marriage. The other situation, in which a man refuses to live with his wife, is rare because, by initiating the procedure of break-up of a marriage, the man stands to lose the bride-price that he may have paid to the wife's natal kin. In a *rajikhushi* marriage, if a man is unhappy, he would desert his wife. If a woman with small children leaves to join another man, she usually has to obtain the consent of her new spouse to keep the children from her previous union. Sukurbati threatened her second marriage partner with suicide if he did not accept her two children from the first marriage. Thus Nimari women are to some extent able to wrest a privileged relationship for themselves.

Absence of Extended Unilineal Descent Groups
The domestic group on Nimari can be viewed both as families and as households; and in the majority of cases a 'household' is also the family unit. The emergence of either an extended unilineal descent group or the socially mobile nuclear family units is conspicuously absent in the 'lines'. Since there is little cash available and the people are almost always penniless, there is little differentiation in the

community on the lines of economic status. An increase in the number of common-law marital unions shifts the focus of authority from the extended family to the conjugal pair. The relative concentration of decision-making power in the conjugal pair has resulted in playing down the authority of a father figure in family relationships, hence a relative absence of family patriarchy in the labour 'lines'. The conventional economics of describing woman as a 'dependant' of man, and the accepted status of man as 'breadwinner' and the household head, does not apply to the living and working patterns of Nimari labourers who do not recognize the category of 'household head' and find little use in their everyday life of the traditional concept of the 'family head' (*mukhiya*). As far as the estate management is concerned, all labour records are maintained on the basis of households and gender relations can be clearly analysed in terms of household organization despite an absence of the concept of household headship revolving around a male.

Division of Labour within the Household
The normal pattern of household activities is based on the premise that both the wife and the husband go out to work because even the unemployed spouse goes out whenever temporary work becomes available. Based on the criteria of a common kitchen and common purse there are 533 households in Nimari labour 'lines'. Households in the size-range of three to six persons are in the majority, and it must be noted that households with large membership are not necessarily with more earners. Only four out of 533 households have four earners. The largest number of households, that is 46.7 per cent, are single-earner households followed by 45.5 per cent households with two earners. In single-earner households, as said before, the unemployed members avail themselves of opportunities for casual labour jobs as and when these are available. Table 7.2 shows ten categories of Nimari households by kinship. There are no households with non-kin members. The most numerous type (57.21 per cent) has a married couple, by either the *bandobast* or the *rajikhushi* type of marriage, and their unmarried children. A unit of wife, husband and children is not only a statistical norm on Nimari but is also the most common type of household. Even the behavioural patterns of persons in other types of households are guided by the norms found in the 'ideal' type.

In 67.22 per cent (i.e. type 3 and 4 in Table 7.2) of households, 43 per cent are supported by the wives only, with the husbands supplementing the household income by taking up temporary work in the tea garden or in some non-estate labouring job. The high visibility of women's active role in economic sustenance cannot be overlooked in our discussion of gender relations.

Gender Roles in Household Organization
Set in the overall context of the plantation economy, the mainly economic content of household relationships seems to have overtaken whatever norms of family

Table 7.2.   Composition of Households in Nimari Labour Lines By Kin

Type of Household	Number of Households	Per Cent
1. Single-member, female	11	2.06
2. Single-member, male	48	9.00
3. Husband and wife	57	10.69
4. Husband, wife and children	305	57.22
5. Husband, two wives and children	5	0.93
6. Husband, wife, children and parent/sibling of husband/wife	28	5.25
7. Truncated 'A' – without father	45	8.44
8. Truncated 'B' – without mother	26	4.87
9. Siblings – without either parent	3	0.56
10. Other types	5	0.93
Total	533	99.95

relationships the labouring community might have had. Setting up of a household takes place at a mature age when both the wife and the husband are expected to be earning members of the community. With the birth of children and the formation of a complete nuclear family, the element of self-reliance on the part of the new household leads the couple to a division of labour. When both parents work, the mothers take their young infants to the tea fields and leave them in a basket by the pathside. Children aged five or more are left at home in the care of older siblings, aged eight or more. At an early age the children of both sexes are introduced to the task of child-minding, and all the female children are also inducted in house-keeping, while the boys learn to fish, grow vegetables and sell fruit from kitchen gardens.

When children are rather small, the woman receives considerable help from her husband in cooking meals and washing clothes. If there are only boys in the family, one of them is taught how to light the fire and cook rice. A boy is generally expected to be an assistant to his father in kitchen-gardening, cattle-rearing and paddy cultivation, while the girl is encouraged to learn the art of plucking tea leaves by helping her mother in the fields. In general, though there is a segregation of roles, frequent crossing of boundaries disavows difference. The boundaries are not fetishized into a spectacle of difference. Again, it is relatively easier for a girl to obtain permanent employment in the garden than it is for a boy because he has to also pass a medical test of fitness. There is no such requirement for female workers. This relates to the planters' perception that work carried out by male workers is hard manual labour requiring, as a prerequisite, medical proof of physical fitness. Female workers, on the other hand, are considered to be involved in light work and therefore not requiring a test of physical fitness. This of course rests on a myth created by the planters for the purposes of categorizing plucking as a non-skilled manual job. A plucker's job requires application of skill in addition to the physical exertion of standing in a position for many hours under the hot sun. C.A. Bruce (1839: 67), a pioneer in the tea industry in India, wrote, 'The plucking of leaves may appear to many a very easy and light employment, but there are not a few of our coolies who would much rather be employed on another job.' Apparently, the exploitative labour process on the garden has devalued women's labour and relatively easier access to a permanent job for a female does not as such enhance her economic value, or assign her a position of superiority in the community. At best, our data can only point to a pattern of relative parity in division of labour in the household organization.

The responsibility for indoor tasks is exclusively female, but female roles are not confined to indoor tasks alone. Here again we find crossing of boundaries of generally perceived male–female tasks. Male members of the household carry out such tasks as growing vegetables in the kitchen garden, tending any livestock that there may be in the family, growing paddy on any plot allotted by the estate, and generally repairing the house and furniture. In cases of the wife's illness, or the presence of newly-born infants, the men do help in cooking meals and washing clothes. For women and men, in almost equal measure, their life is a perpetual cycle of work and weekly pay. Their horizons hardly extend beyond the functions of making survival possible. Early socialization of both the female and male workers, in their respective roles in work and home spheres, prepares them for the future assumption of these roles in their adult life. Role-crossing performed by adults, observed as well as emulated by junior members, prepares them for a larger degree of co-operation in the household tasks. There is a parallel role-crossing in the work situation as well. Depending on the production needs over the year, women are expected to work as field workers during the non-flushing season, and men to work as pluckers in the flushing season. Thus, in practice, role-crossing is fairly common among the workers both in 'on' and 'off' work situations. Non-discrimination is also borne out by the fact that Nimari parents do not display any preference for one sex or the other; they celebrate the birth of any child. For similar data on tea gardens of the Dooars in North Bengal see Bhowmik (1981: 127).

Relative parity in gender relations on Nimari has, however, not meant a life of relative comfort or advantages to women. Compared to men, supervision of women's work in the tea fields is more closely monitored lest a careless plucker resorts to either too fine or too coarse a pattern of picking and thereby brings about a variation in the quality of tea obtained and manufactured in that garden. Women have to not only be particular about their performance, they have also to tie their work to a more rigorous time-schedule. Unlike the male workers who nearly always finish their tasks early and leave for the 'lines' by 2.00 p.m., the female workers have to remain away from home for at least eight hours each day. Also, being a slow plucker is a clear disadvantage to a woman because in the flushing season she is unable to earn extra money while a slow male field worker suffers no such disadvantage. Even if he takes longer to finish his task he earns just the same as a fast worker. A woman worker, on the other hand, not only spends longer hours at her work than a man does, she has also to try and increase her speed to earn better wages. In this way there is clearly greater pressure on the woman to increase the productivity of the enterprise.

As far as the occupational mobility of the women workers is concerned, all able-bodied women are offered the job of plucking and only those found plucking extremely slowly or making errors are asked to take up field work. Being a field worker means a lesser chance to earn extra wages in the flushing season. For this reason no woman on Nimari would by choice become a field worker. The other occupations, for example those of factory work and supervision, are offered to men only. Women themselves do not aspire to these jobs as they are socialized to consider themselves incapable of coping with the requirements of handling machines or commanding the gangs of labourers. The tea companies do not consider women to be fit for such jobs, hence the workers' internalization of gender-related stereo-types of what women can and cannot do. There are hardly any jobs available to women outside the gardens, with the exception of possibilities of occasional offers of transplanting paddy on nearby farms. Nimari women have access to only those jobs which require both hard physical labour and long hours.

The Nimari men also engage in tasks which require equally hard labour. They put in longer hours in order to earn overtime and bring their earnings parallel to those of the pluckers (females). On rare occasions, some women after childbirth find it hard to resume their normal occupation of plucking. They are sent to the sorting department of the tea factory to separate the grades of manufactured tea. Such women are not considered factory workers and they know that they are there only for a short period.

It is clear that though the tea garden women have a large measure of visibility within the plantation milieu, their visibility is no special advantage to them in terms of social status. At the same time compared to plantation women workers in the Caribbean and the Pacific where women were given tasks of field work, the tea garden women perform a relatively more skilled job and thereby deprive men, so to speak, of their special status. This lack of special standing of the men in the work situation, when carried forward to the area of gender relations in the community, results in the weakening of the patriarchal family system. In the absence of extensive patrilineages patriarchy is not emphasized on the tea gardens of Assam. Rather, because of the fragmentation of lineages between the village and the gardens, much depends on the particular circumstances of an individual and her or his network of support. Sukurbati and women like her are able to live on their own mainly because they continue to receive support from their families of orientation on the estate. Thus the proverbial break 'between being a daughter and sister in one family and becoming a wife and daughter-in-law in another' (Mandelbaum 1970: 82) is neither sharp nor difficult in the case of Nimari women.

Conclusion

Social relations of production on Nimari reflect the nature of gender-related status differences and at the same time, the socio-cultural features of the tribal and caste groups in the 'lines' also affect the specific nature of gender roles in the community. A fair degree of sharing and role-crossing between sexes does not however imply that their poverty-related problems caused by inadequate working and living conditions do not require immediate attention. Horizontal levels of gender-parity among the tea garden workers in the context of extreme types of vertical inequality have severe limitations and cannot be regarded in any sense as a resource for bringing about change in their class position. The ideology of equality is certainly reinforced by joint consumption of rice-beer (la-pani) by the workers. Most workers brew rice-beer in their homes, and it is common for wife and husband to take a bowlful each in the evening after returning from work (Jain 1988: 51). The sale of alcoholic drinks is banned in the 'lines' but all the same it is sold under cover on Sunday nights. Leaving aside some confirmed alcoholics, most women and men on Nimari take home-brewed rice-beer. In the case of alcoholics, their entire wages are spent on liquor while their households are run on the earnings of the remaining working members. On the basis of a few cases – 2 per cent of the total number of households in the 'lines' – it can not be said that in general most workers on the tea gardens are drinkers, with their womenfolk running the households on their incomes alone, even though this widespread belief is prevalent among the tea garden managerial and supervisory staff. My study of household budgets in the 'lines' shows, in fact, that the labouring families are able to survive only when there is pooling of incomes of both women and men to meet the household expenses (see Jain 1982: 58).

Scholars of colonial plantation labour may find a mismatch between the above account of gender relations on Nimari and most general accounts of labour exploitation and of women's oppression in particular. The foregoing account of gender relations is, I fear, likely to create (in fact *has* created in some quarters, see for example Kannabiran 1989: 370) an uneasiness and is therefore unacceptable to those feminists who advocate the politics of irreducible differences. Preoccupied with binary oppositions, they seem to have built a discourse around a series of dichotomies (public/private, nature/culture, production/reproduction and so on), making them as hegemonic and universalist as are the traditional power structures they want to change. Feminist critics often show a great deal of intolerance for new ethnography and new voices. If a study moves beyond the exclusive focus on differences and speaks of hybrid realities which are experienced everyday by both women and men, it is perceived as premised on a unique situation, and is ridiculed as a miracle without any location in historical or contemporary sources. One wonders why there is so much fetishization and celebration of differences. Undoubtedly traditional sex roles frame most social structures, yet endless variations within them do pose a question regarding the rigidity of such models (for recognition of enormous variability see Leacock 1986: 128–9; Uberoi 1987: 388–90). I have ventured to deal with shades of grey between black and white. Gender relations in this labouring community on Nimari are glossed with loss of patriarchal authority. The end product seems to be a degree of parity in the day-to-day life of the workers. This does not mean that in order to gain equity one has to be poor. I only wish that the reverse were true and all the affluent, powerful and educated women had access to equality. But that is another world of dilemmas and multifocality.

Notes

1. In Northern India, the tea plantations are called 'gardens' and in Sri Lanka, Kenya and South India 'estates'. In this chapter, the terms 'garden,' 'estate' and 'plantation' are used interchangeably to denote an agro-industrial unit producing and manufacturing tea as a saleable commodity, specially for world markets.
2. 'Nimari' is the pseudonym for the tea garden on which the author lived to carry out anthropological fieldwork during 1978–9. I have used pseudonyms for all names of places and persons in the region.
3. 'Home brew' refers to the rice-beer prepared by the residents of Nimari labour lines by mixing paddy and yeast and allowing it to brew for seven to ten days. Rice-beer is taken by both women and men as an invigorating beverage on almost a daily basis after returning from their respective tasks. During life-cycle rituals it is offered to guests as a mark of respect and welcome.
4. On Assam tea plantations the labour quarters are known as labour 'lines'.
5. Of 1,555 dependants out of 2,076 persons living in the labour 'lines', 37.56 per cent in the thirteen to fifty-three age group seek temporary employment on Nimari. Besides them, depending on the estate's demand for labour at different times of the year, many more temporary hands come to work on Nimari from the nearby 'bustee'. These are clusters of hutments of ex-tea garden labourers, who procured small plots of land on the peripheries of the tea gardens and settled as small farmers after the expiry of their contracts. Because of the small size of their plots and the poor fertility of the soil, the bustee-dwellers have to depend on the tea garden employment to make their survival

possible. The planters encourage the retiring labourers to settle around the estates to ensure the availability of extra labour when it is required.

6. In contrast, on rubber plantations in Malaysia, if a female worker becomes a widow or a divorcee or a separated person, she is expected to return to her natal home which is often in another estate. The estate does not allocate a house in the name of its female workers (see Jain, R.K. 1970: 85–6).

7. For a comparison between sardar on Assam tea gardens and *kangani* on Malaysian rubber estates and Sri Lankan tea estates see Introduction (p. 12).

8. In Assam, people living on the tea gardens and those connected with them keep two clocks – one following the Indian Standard Time (I.S.T.) while the other shows the 'Garden Time', which is an hour earlier than I.S.T. All work and off-work activities on the garden are regulated by 'Garden Time'.

9. In 1978–9, Rs. 8 to 9 were approximately equal to US$ 1.00. The wage hike, per the Assam government order of 1990, resulted in an overall increase in the cost of tea production. The minimum rates of wage for adults and non-adults were fixed for specific zones and for three successive years. In all cases the first year was taken as the period between November 1989 and 30 October 1990. By the third year, i.e. from November 1991 to October 1992, in zone A (where Nimari tea garden is located), the minimum wages for an adult rose to Rs. 17.60.

10. Considered to be a part of the workers' wages, the benefits include the free housing, medical care, subsidized rations of rice and wheat flour, sick leave, maternity benefits to women workers, free firewood, 600 grams of tea per year per family, protective material like a blanket, sandals, tarpaulin sheet and an umbrella, for use during the work. In theory, the estate also provides creches, creche-attendants and milk for children up to two years of age and solid food for children between three and six. In practice, the creches are run inefficiently and the working mothers have already evolved their own mechanisms to cope with child-minding.

8

Women's Role in the Household Survival of the Rural Poor: The Case of the Sugar-cane Workers in Negros Occidental

Violeta Lopez-Gonzaga

Overview: Women in Philippine Society

The Filipinas have traditionally occupied a strategic role in Philippine society. From pre-Hispanic times, a Filipina's right to legal equality and to inheritance of family property have been an accepted norm. Contemporary Philippine society continues to reflect the strategic role played by the Filipinas in the shaping of their history. In fact, across the different social classes, a commonality present among all Filipinas is the critical role they play in household management. Under crisis, this role has become more accentuated, especially among the poor. Previous studies among rural proletariats and peasants in Negros Island, the sugar bowl of the Philippines, have shown that under crisis women have assumed the major task of ensuring their household's survival. Where their men have been rendered jobless and virtually helpless in the collapse of the island's main industry, the Negrense women have ingeniously assumed sundry jobs and petty forms of entrepreneurship to ensure their family's daily subsistence. In the cities, the same holds. Working as laundry women, peddling sundry wares and foodstuff, or working as barmaids and hostesses, even going to the extent of peddling themselves to ensure their household's survival, Filipinas have shown tremendous resilience amidst absolute poverty.

'Crisis' is defined in this chapter as 'an unstable state of affairs in which a decisive change is impending – a psychological or social condition characterized by unusual instability caused by excessive stress, requiring the transformation of existing cultural patterns and values' (adapted from Webster Third New International Dictionary 1971). It is under such circumstances, particularly in conditions of insurgency, that I would like to examine the critical role played by hacienda-based *duma-an* (i.e. plantation resident sugar-cane women workers) in their household survival.

The Historical Context and Socio-Political Factors

A study of the history of Negros shows that the contemporary crisis is something closely intertwined with capital expansion and the rise of a monocrop economy in the island at the turn of the century. This one development, ushering in a new economic order (Lopez-Gonzaga 1987), cannot be divorced from the broader realm of international economic relations in the mid-nineteenth century – a period marked by an unprecedented level of technological achievement for Western Europe.

International economic development in the mid-nineteenth century came to mean, among other things, a relatively steady increase in the consumption of sugar – a commodity previously enjoyed only by the elite. With the trickling down to the working class of the economic benefits of the technological revolution, especially in the United Kingdom, sugar became accessible to a wider number of people. As documented by Mintz (1985: 197), between 1750 and 1850 sugar ceased to be a luxury and became a necessity. Thus, between the 1850s and the first half of the twentieth century sugar became an important commodity in international trading.

By the time the Philippine Republic was established the sugar industry had already gained a reputation as the elite among all the Philippine agricultural-based industries. It had also become the main basis of wealth and powerholding of a strategic grouping of landed gentry – the hacienderos, now commonly referred to as the sugar-cane planters. Because of its vast capitalization and large labour force (431,000 in 1976), historically sugar had been the most profitable and powerful of the export industries (McCoy 1983: 1). Until the major sugar crisis of the 1980s, the industry was a significant source of the country's foreign exchange. In fact, sugar generally accounted for 25 per cent of the total exports (NEDA 1982: 345–55). When in 1978 sugar exports went down to only 6.3 per cent of total exports, the hacienderos were shorn of their powers.

Before the worsening crisis, and as late as crop year 1982/3, Negros Occidental province had the major share of the Philippines' raw sugar production and 44 per cent of the national sugar hectarage. This wholesale dependence of the province on sugar production – sugar being its very lifeblood – led to massive labour displacement, the widespread incidence of absolute poverty, and consequently hunger. After decades of exploitation and excesses by the once powerful hacienderos and the collapse of sugar prices in 1983 and the contraction of the market, the masses of displaced peasants and unemployed workers became restive. Life for those who stayed in the sugar-cane farms has not been any better than for those who moved out into the city. Completely dependent on the low wages from their labour in the farms, many were not paid and were eventually displaced when the planters stopped receiving payments for their sugar. With no other options for work, those in abandoned sugar-cane farms simply faced up to reality, readjusting their meals to include edible plants, field rats and wild root crops.

The Household

A common characteristic of the poor is the dominant role played by the household in their survival. A recent study conducted among the poor in Mexico strongly supports this view (Selby et al. 1987: 419). The researchers found that a key strategy for survival among the poor is organizing into an 'efficient household' – a large extended family with many children, with either distant relatives or grandchildren present, the older ones in the workforce, and a household head with a steady job (Selby et al. 1987: 420). This finding presents a clear contrast to the consistent survey findings of this writer and the Bacolod Social Research Centre (SRC) research team findings which reveal the predominance of the nuclear, rather than the extended structure of the family. Traditional rural Filipino households were of the 'extended household' type. However, the drastic rise in the number of dis-possessed peasants seems to have forced a shift to the nuclear type of household organization. With no more land to till, rural Filipino folk may have found the nuclear type of household more suited to their survival strategies. In a very unstable state of affairs, such as the *gamo* or financial pressure, a smaller household size became imperative for household survival.

Among the sugar-cane workers, there was also the observable trend of the 'segmentary household' or single parent households resulting from either break-up (due to marital estrangement) or forced relocation of household members (due to *gamo*). For example, a spouse may have been forced to work in Manila for household survival. By Phase I in the SRC study (Lopez-Gonzaga 1986), it was found that the persistent problem of underemployment and joblessness, especially among some youth in the observed site of the study, triggered an outward migration into the cities. Of the nineteen households with young adults, 68.4 per cent had young members working outside the farm. Among those who went looking for job opportunities outside the hacienda, 79 per cent ended up being domestic helpers in Manila and La Castellana. Two with special training landed jobs as driver/ mechanics, while one migrated overseas as an 'entertainer'.

While some young adults (whose ages range from fourteen to twenty-six) have gone out, there are also a number who have returned home, disappointed about the extremely low wages they were paid at their points of destination. Having worked long hours in strenuous jobs, with low wages, those who returned home commonly complained about the impossibility of saving and sending remittances to their families. Yet parents, desirous of lessening their financial burden at home, encouraged such returning youngsters to hold on to their jobs. This adaptive strategy for dealing with burgeoning household expenditures may be inferred from the case of Ruby:

> Ruby is the fifteen year-old daughter of Tio Mocro and Tia Lilia. Despite the fact that both her parents are working, she has long desired to help her family by seeking work outside the hacienda. This is because she has seen very little chance of being given work by the hacienda. Also, she desires to pursue the 'promising life in the city'. During the first quarter of 1984 she sought employment with her friend's *amo* (mistress) at a monthly salary of two hundred and fifty pesos. With this salary she managed to send, every two to three months, the amount of one hundred pesos, and at times a lesser sum. In Holy Week, 1986, she went home with *pasalubong* (gifts given on arrival) of bread and simple clothes for her parents and the younger siblings. During her stay, however, she seriously thought of just remaining there because of her dissatisfaction with her *amo*, her low salary, and her occasional feelings of homesickness. Yet her mother convinced her of the great need for her to go on working if only to support herself, *bisan para nalang sa kaugalingon mo nga lawas* (for you to have a source of income even just to support yourself).

A key factor affecting the relative flexibility of the sugar-cane workers under crisis i.e. to adapt and survive, is their household size. Yearly surveys (of sugar-cane workers' communities) conducted by the SRC since 1982/3, have consistently shown the comparatively large size of their households – the mean size being six. With a low level of educational attainment and low earning power, the sugar-cane workers manage their households poorly. Generally, they find it difficult to shape their destiny. For displaced workers who are not organized, their fate seems confined only to bearing and suffering crisis conditions, as may be seen in the case of Tyo Luis:

> Tyo Luis, 48 years of age, is one of the displaced sugarcane workers of Hda, Sta. Monica. Since his displacement only one son works full-time as a fisherman, and is, in fact, the sole breadwinner of the household. His day's catch is reserved primarily for the household's subsistence. Any surplus is sold in the market to generate cash. More often than not, however, the catch is insufficient to meet the household requirement. This Tyo Luis attributes to the dynamic fishing activities of some fishermen from the neighbouring *barangay* (smallest political unit of the Philippine society, equivalent to the *barrioor pueblo*). Some household members occasionally engage in charcoal making. Their main problem, however, is the distance between their place and the market.
>
> Apart from himself, the household of Tyo Luis is composed of his wife and ten children. (Bañas 1986: 63)

As may be inferred from the case of Tyo Luis, the question of a cane worker's relative flexibility under crisis is also determined by other factors, like the number of working household members, the dependency burden, the skills of members who are of working age, and the social network system. It is worthy of note that while other household members can produce marketable commodities such as charcoal, the absence of a trading network deters them from further developing this alternative source of income. The absence of social connections – a characteristic of households in conditions of absolute poverty (Lopez-Gonzaga 1985) – deprives them of access to even basic social services.

Based on the survey findings on the above mentioned factors which may determine household flexibility, it may be concluded that the sugar-cane workers are grossly limited in their ability to deal effectively with the crisis. For instance, the 1985/6 data show that the most commonly mentioned skills of the respondents are fishing (54 per cent) and animal husbandry (52 per cent). Since these skills are both basic and agri-based, the displaced workers' employment opportunities are severely limited (Bañas 1986: 18). In fact, only 16 per cent of the migrant workers were employed either under a *pakyaw* (piece rate) arrangement or on an irregular basis (Bañas 1986).

Household Income, Socio-Economic Activities and Women's Contribution

The 1987 survey data show the same high rate of unemployment and under-employment of the sugar-cane workers. Data on the number of working household members consistently show a median of only one member working (the mean for 1987 is 1.62). In terms of income a slight increase was evident in the 1987 survey finding. A comparative study over a five-year period of the household income of the sugar-cane workers shows that between the milling season 1982/3 and August 1987, in real terms, the monthly income of the workers increased by only 12 per cent (Table 8.1). Between 1984/5 and 1985/6, a noticeable drop in the monthly household income at current and constant prices was recorded. The drop in the income and purchasing power of the workers between these years was as high as 51.9 per cent. This finding also shows that the economic crisis was at its worst during the milling season 1985/6 – the last year of the corrupt Marcos regime.

A further analysis of the growth trend of the household income of the workers shows a positive course. When deflated by the regional Consumer Price Index (CPI), wages between 1985/6 and the time of the 1987 survey are revealed to have risen in real terms, by as much as 36 per cent (Table 8.2). Still, the median monthly income of 792 pesos in 1987 is 49.5 per cent below the estimated food threshold

Table 8.1. 1 Monthly Income of Sugar Workers, at Current and Constant Prices, Negros Occidental, 1982–1987

		Monthly Income	
Data Source	Year of Study	Current	Constant*
Gonzaga (1983:17, Table 14)	1982/3	320	184.44
Gonzaga (1985:100, Table 8)	1984/5	603	203.78
Bañas (1986: Table 11)	1985/6	503	134.10
Gonzaga (1987: SRC/VRC-Ford Study)	1987	792	209.80

* Current Prices Deflated by the Region VI Consumer Price Index (CPI), with 1978 as Base Year.

Table 8.2. Consumer Price Index (CPI) for all Income Households, by Geographic Area (1978 = 100)

	CPI (all items)		
Year	Philippines	Western Visayas	Negros Occidental
1981	157.1	158.3	155.9
1982	173.2	173.5	169.7
1983	190.6	188.8	180.0
1984	294.8	295.9	296.9
1985	352.6	375.1	380.2
1986	355.3	373.4	375.9
1987	365.4	377.5	380.2

or basic subsistence level for a family of six members (at 1,569 pesos per month for the Western Visayas area for 1985) (Mangahas 1985).

What it means for the surveyed families to fall below the poverty threshold i.e. in terms of household budgeting, diet and over-all nutritional status may be inferred from the following synopsis of a case study of a sugar-cane family living in a hacienda in 1985/6:

The Gaitano family was one of the worse off of type C households (an arbitrary classification of extremely impoverished sugar-cane workers' households). I observed it closely for six months. The family was composed of three female and three male children, the youngest of whom was one year old. The mother, Lilia, had been forced to help her parents earn additional income by working in the sugar-cane fields herself. Faced with no better prospects, she married four years later. She indicated that life was tolerable during the early years of her marriage. Work was readily available, and she had only three children for whom she and her husband were then able to provide adequately. Even though she was breastfeeding them, she managed to buy them condensed milk and vitamin supplements. She lamented how deprived her three younger children were now, compared with the first three. With her husband's twenty pesos per day income, and an average of four working days per week, she found it impossible to buy milk, or even just simple nutritious food for her children. Lilia said she was rarely able to prepare a completely nutritious *laswa* (a vegetable dish prepared with dry shrimp or shrimp paste) when they had money. Their daily *laswa* now consisted of just boiled edible foraged leaves and salt. Other days the family would subsist on salted or dried fish. Three pieces of small dried *sap-sap* cost one peso. Usually, Lilia bought three pesos worth of *sap-sap* which she allocated for two meals. Each family member would then get a tiny ration of the dried fish, usually boiled and cut into halves to make it last for the next meal. (Da-anoy 1986a)

As may be deduced from the case study of the Gaitano family, the wife plays a very critical role in the 'pliancy' or adaptability of a household undergoing financial crisis. Working with a tight budget the mother of a one-member-working household is under pressure to maximize total household income. Often the budget-stretching act of a mother is only possible with the parallel action of 'belt-tightening' among other members of the household. Obviously, the Gaitano family is not a 'well-managed' one, as may be seen in their great dependence on purchased food at the local stores where the price of commodities is usually doubled. The household has no food storage facilities which would allow them to store foodstuffs which are seasonally available in relative abundance. Therefore they are greatly dependent on the market for their subsistence, and subject to sudden increases in the price of commodities. Thus, though some workers were able to produce rice from borrowed lands, the Gaitanos' household economy remained tied to the market forces.

Close observation of household has, in fact, shown that apart from serving as their basic staple food, rice is also used by indebted worker households to pay their debts. It is used in bartering for fish with a *pantingero* (fish vendor). In times of emergency their rice produce is sold at very low prices to the *manogdalawat* (rice buyer from the town). The wide range of uses for *palay* (rice grain) and rice as a household mechanism for coping with financial crisis is as follows:

a) *Padalawat*. A system which involves the housewife's selling a portion of her *consumo* (rice stock) at only ten pesos per *ganta*. She actually loses for every *ganta* which she sells out. This practice, however, enables her to purchase the commodities which she needs for home consumption every time she falls short of cash.

b) *Baylo* or barter. In this type of economic activity, a housewife may accept fish from her *pantingero* (fish vendor) in exchange for a portion of her rice stock or *consumo*. She may also offer it to the local store owner in exchange for soap, sugar, kerosene and other commodities. Often, a housewife draws an advance supply of these commodities, and pays for it when she receives her *consumo* from the hacienda.

c) *Pangalini*. With this system, a housewife takes a sum of money as a loan from her regular credit source (moneylender or store-keeper). This is to be repaid during the harvest season. The repayment is made, not in cash, but in the form of *palay* or grain. The *palay* costs 90 to 100 pesos per cavan (equivalent to one sack of rice). For every 50 pesos she borrows, she agrees to pay one *cavan* of *palay*. By this transaction, the borrower actually loses 40 to 50 per cent of the actual cost of the *palay*. The money borrowed is used either to buy fertilizers or to pay for the rental of *carabao* (buffalo) for ploughing their portion of the rice field, or else, to cover emergency expenses, as in the case of accident or illness.

d) *Panaghaw*. This practice allows a single-income-earner household to glean excess grain after the *palay* has been harvested by the major workers. This activity, usually undertaken by the women, is commonly resorted to by the squatters within the hacienda.

e) *Pamupho*. This system allows a household with one or no income earner to gather rice which has been spilled during threshing. Again, this is usually resorted to by the poorest households in the hacienda – the squatters.

Apart from these additional economic activities to cope with financial stress, there are other adaptive strategies to which women resort:

a) *Prenda* is the pawning of goods such as a television set, sewing machine, or radio, or of the title of ownership of a *carabao*. This is the mechanism commonly resorted to by the workers of better-off households, who possess moveable household goods. The pawnbroker and the pawner agree upon the rate of interest and the date of repayment. Failure to meet the conditions of the agreement means the outright confiscation of the pawned property. As for other coping strategies, *prenda* is resorted to in cases of emergency or when schools are due to reopen.

b) *Saka-an* or 'five-six' is a system which allows a housewife to avail herself of a cash loan at an interest of 15 to 20 per cent per month. With this high rate of interest, *saka-an* is often the last resort of households in deep distress.

c) *Pasagod* is the most beneficial of all the adjustment mechanisms of the workers' households. This is the local term for swine, *carabao* (buffalo) or poultry husbandry. Usually, a trusted neighbour is asked to raise or fatten animals which are owned by better-off worker households. The cost of feeds is often shouldered by the one who fattens the animal. When the livestock is ready to be sold, the person who raised it consults the owner about its selling price. After the sale, the proceeds are divided equally between the owner and the one who did the rearing.

It is significant that the main response to pressing problems of financial stress is to work harder. This is clearly exemplified in the workers' subsidiary economic activities. An overwhelming 96 per cent of the survey respondents raised animals, mostly pigs and chickens. As previous SRC studies have shown, the livestock and poultry holdings of the sugar-cane workers represent their only capital holding. During times of dire need or times of emergency, their animals provide them with the means of generating quick cash. In addition to raising livestock, all respondents reported their households as cultivating their own crops, the most commonly produced being vegetables, grown by 94.6 per cent of the workers. The observation data show that privately grown vegetables are mainly for household consumption. Some more enterprising households, however, generate a surplus which they sell seasonally to gain the much needed cash for their basic commodity needs such as kerosene, salt, soap, cooking oil, and so forth. Per harvest, the mean amount of cash generated from their agricultural produce is approximately 40 pesos. In some cases, a type of 'social insurance' provision is observable. Instead of selling their surplus to the market, it is distributed to their recognized *kasilingan* (regular exchange partners) who may be relatives or neighbours. The shared crop is, of course, viewed as some kind of 'deposit' for a future expected return gift from the *kasilingan*. Other than vegetables, the respondent households also plant rootcrops, corn, beans, herbs, and even decorative plants (grown by 40 per cent of the workers). Apart from the desire to beautify their home surroundings, some of the more enterprising respondents make some money out of their decorative plants by selling them (especially on Fiesta Minatay or All Souls' Day). A significant minority (34.6 per cent) grow herbs – a probable indication of the inroads made by some private volunteer groups' attempts to lessen the poor people's dependence on imported drugs.

As delineated in the description of the different adaptive strategies of the workers, some measures, resorted to simply because of a lack of better alternatives, have negative effects on their overall economic status. A good example of this is the *saka-an* or 'five-six' which encourages the accumulation of debt because of the high usurious rates. Other adaptive strategies are only seasonally feasible, and allow only a temporary alleviation of the difficulties of survival. The long-term view of security is an elusive factor in their overall adaptive strategies for household survival. In the 1985/6 SRC-NEDA study of the status of displaced sugar-cane workers, Bañas (1986: 90) found that the main objective of this category of workers 'is to forestall their physical annihilation rather than the maximization of their satisfaction'. The displaced workers, according to Bañas, attempt to improve their income 'by maximizing their only major resource, their labour'. The income maximization strategies of displaced workers as identified by Bañas are as follows: (1) to continue working in the sugar industry but in a different locality, as exemplified by the migrants; (2) to shift to other sources of livelihood; (3) to increase their number of occupations and/or working hours; and (4) to induce, if not compel, the other household members to engage in productive activities.

Time Allocation, Household Division of Labour, and the Contribution of Women and Children

A question posed with regard to household division of labour and the amount of time spent for each work activity reveals some interesting patterns. The survey finding shows that the mothers bear the greater responsibility for such domestic chores as cooking, cleaning the house, washing clothes, and watching small children (Table 8.3). Only a very small percentage of the husbands share in such culturally-defined 'female' responsibilities. Yet, in traditionally-defined 'male' functions such as farming, as many as 14.5 per cent of the mothers were reported to be actually engaged in earning a living for the household. In fact, a previous study conducted by this researcher, showed that under the worsening crisis in 1984/5, women showed a greater capacity for creating self-generated types of employment and non-traditional sources of income, such as gathering *ipil-ipil* leaves for animal feed (Lopez-Gonzaga 1985: 33–5). A later report by this writer reveals that women in lower-income Negrense households are also the providers of food, fuel (89 per cent of the surveyed respondents use wood for cooking), water, and part or all of the family income (Lopez-Gonzaga 1987: 24). The pressure is greater for low-income women who are single parents. Already burdened with a large household, a widow or a single parent becomes hard-pressed when forced by circumstances to combine wage earning and child care.

The disaggregated (by sex) employment data, show that the majority of the household-earning members are males. The range of employment of the different household members shows that the majority of the women who are employed (14.5 mothers and 9.6 daughters) work as sugar-cane workers. Only half of the employed female household members have regular employment status. The computed mean number of working hours per week by both sexes, reveals their gross under-employment: the mean number of hours for the males is 41.1 hours, for the females it is 41.6 hours. Overall, the percentage contribution of the females to household income is 18.4 per cent.

It is important to note that the second ranking form of time investment after activities directly related to cash and food production, is baby-sitting. In a study on child-care and rearing practices in Negros Occidental (Lopez-Gonzaga 1987), it was found that one of the major hindrances to mothers' becoming more productive is the presence of babies and toddlers in their families. With the dominance of the nuclear family structure, the traditional support systems provided by extended family relations (such as grandmothers watching toddlers and babies) was not available

Table 8.3. Household Activities, by Member Involved, by Hours Spent. SRC Survey 1987

Variable	Traditional Cane Farm		Abandoned Cane Farm		Transitional Cane Farm		Overall Total	
Household Member involved in:		per cent		per cent		per cent		per cent
Fetching	14		12		38		64	
Father	8	57.14	6	50.00	19	50.00	33	51.56
Mother	5	35.71	4	33.33	10	26.32	19	29.69
Son	6	42.85	6	50.00	24	63.16	36	56.25
Daughter	0	0.00	1	8.33	8	21.05	9	14.06
Other	0	0.00	1	8.33	3	7.89	4	6.25
Cooking	23		12		40		75	
Father	3	13.04	2	16.67	3	7.50	8	10.67
Mother	21	91.30	11	91.67	37	92.50	69	92.00
Son	0	0.00	0	0.00	0	0.00	0	0.00
Daughter	3	13.04	0	0.00	5	12.50	8	10.67
Other	1	4.35	0	0.00	3	7.50	4	5.33
Cleaning	23		12		40		75	
Father	2	8.69	0	0.00	0	0.00	2	2.66
Mother	22	95.65	8	66.67	32	80.00	62	82.67
Son	0	0.00	0	0.00	1	2.50	1	1.33
Daughter	5	21.74	3	25.00	11	27.50	19	5.33
Other	1	4.35	2	16.67	1	2.50	4	5.33
Washing	23		12		40		75	
Father	1	4.35	0	0.00	0	0.00	1	1.33
Mother	22	95.65	10	83.33	36	90.00	68	90.67
Son	0	0.00	0	0.00	1	2.50	1	1.33
Daughter	3	13.04	2	16.67	5	12.50	10	13.33
Other	2	8.69	2	16.67	3	7.50	7	9.33
Farming	22		12		39		73	
Father	17	77.27	11	91.67	32	82.05	60	82.19
Mother	11	50.00	2	16.67	5	12.82	18	24.66
Son	2	9.09	0	0.00	5	12.82	7	9.59
Daughter	0	0.00	0	0.00	3	7.69	3	4.11
Other	2	9.09	1	8.33	3	7.69	6	8.22

A closer study of the overall time use of the respondent households reveals that only 22.2 per cent of this resource goes to direct cash-generating activities, and 19.6 per cent to farming (Table 8.4). This finding actually corroborates their reported under-employment and joblessness.

Table 8.4. Time Allocation per Week, by Activities, SRC Survey 1987

Household Activities	Traditional Cane Farm		Abandoned Cane Farm		Transitional Cane Farm		Overall Total	
		per cent		per cent		per cent		per cent
Fetching Water	7.19	4.59	5.47	3.98	2.84	1.72	4.33	2.74
Cooking	17.41	11.12	9.73	7.07	16.57	10.01	15.72	9.95
Cleaning House	6.54	4.18	6.03	4.37	6.58	3.97	6.48	4.09
Washing clothes	9.96	6.36	9.50	6.91	10.11	6.11	9.96	6.30
Farming	26.41	16.87	23.08	16.79	36.32	21.93	31.01	19.62
Earning a living	35.64	22.77	30.50	22.19	36.35	21.95	35.14	22.23
Baby sitting	53.36	34.09	53.17	38.67	56.82	34.31	55.42	35.06
Total	156.51	99.98	137.48	99.98	165.59	100.00	158.06	99.99

to the mothers. Since the type of employment available to women generally falls within the 'informal' sector category, the opportunities for work are usually found outside their homes (Lopez-Gonzaga 1987: 35). The range of employment available to them, as revealed by the survey conducted for this study, are sugar-cane field work, rice and sundry farming, laundering, and domestic service. Thus, the outside jobs available to the mothers are forms of economic activity which are incompatible with simultaneous child-rearing. The following case of Lilia Gaitano, as recorded by Da-anoy (1986a), delineates well the difficulties of a mother with a baby, toddler and other young children.

> It was around 5.00 a.m. when Tio Gusto got up from his bed to fetch water. He took the five-gallon plastic container and went to the nearest *bumba* or pump well which is about a kilometer's walk away from their house. This activity took him about twenty minutes to complete. Tia Lilia was awakened by the noise made by Tio Gusto's movements. She yawned and stood up, took a look at her baby sleeping beside her two other sons, and asked Nene, her daughter, aged nine years, to get up and cook. The rest of the children were still fast asleep. Nene started frying the *bahaw* or the left-over rice from the previous night's meal. After this, she boiled an additional tuna can of rice. Between 5.25 and 6.35 a.m., Nene cooked and washed the dirty dishes from the previous night's meal. Finished cooking, Nene took their new porcelain plates, filled each with fried rice, and placed them on the floor where the children usually sit to eat. The children, including Nene, ate

the *kalo-kalo* with relish, though they had no meat. In less than ten minutes, everyone had finished the meal.

At about 7.00 o'clock, Nene prepared herself for school and soon left the house. The floor on which they had eaten was left unswept. The house dog ate the morsels of food on the floor, literally cleaning it up with his tongue. As usual, the house was left messy after all the morning rush.

In the case of working mothers, the previously referred to 1987 study of this writer, shows that they use substitutes for child-care and feeding when they go out to work (Lopez-Gonzaga 1987: 36). The fact that their sons and daughters are the substitutes frequently identified by the mothers, however, poses questions about the quality of child-care which their younger children are receiving. Further data analysis of the survey findings for the same study reveals that for every two younger children ranging in age from zero to six years, there is approximately one child within the category of seven to fourteen years who may be able to care for the younger siblings during their mother's absence.

Observation data show that a common practice of working mothers is to boil an extra amount of rice to free the baby-sitting child of the burden of cooking. During the daytime, therefore, the children of working mothers, subsist mainly on *bahaw* (cold left-over rice) with salt or sugar as 'relish'. The alternative to this staple diet is food equally low in nutritional value, like coloured chips, galletas, biscuits and other flour-based products.

Though not often eager workers, older children, especially girls eight years old and above, provide a measure of relief from the heavy domestic load of mothers of poor Negrense households (Lopez-Gonzaga 1987: 36). In fact, mothers depend on their older children, especially daughters, to take care of cooking, and feeding their younger siblings as may be seen from the following excerpt of Da-anoy's (1986b) field notes:

Nene left the house without combing her hair since she had lost the family's comb. Tia Lilia scolded her for doing so. Meanwhile, the youngest girl of the family urinated on the floor, adding to the mess in the house. Tia Lilia angrily took the rag and wiped the wet portion of the floor, at the same time taking the child from the floor to her hip. She started cleaning the floor, then washed the dishes and bathed the three younger children. She commonly made unpleasant remarks about her son, Timoy, who often refused to take a bath. This done, she dressed them in not too clean or thoroughly washed clothes, and then rested on the balcony. That was about 8.35 a.m. At about 9.00 a.m., she breastfed her baby while the two pre-school children played in the yard. Nonoy, the eldest of the boys, aged eight years, also joined these two, since he had refused to go to school. The meagre *baon* (snack money) given to him had made him sulk and decide to stay home. Tia Lilia was supposed to do their laundry at 10.00 a.m., but she postponed it for the next day. During these hours, 9.00–11.00 a.m. Tia Lilia is usually free from household chores. At this time, she reclined on the balcony with her little baby girl. When Nene returned from school at 11.15 a.m., she asked her to do the cooking. On Nene's refusal, reasoning that she was tired, Tia Lilia gave the baby girl to her and cooked the rice herself. She did not worry about their lack of meat. The children were used to eating just rice with salt. Occasionally, they were given fish or vegetables. They have been conditioned to eating rice with ginamos for meat. The whole family ate together upon the arrival of Tio Gusto. Each plate was filled with rice. As in the morning, they had nothing to go with it, only salt. They ate very fast, like very hungry individuals. I could even hear the sound they made as they chewed their food. They invited me to eat with them. The children ate naturally, unmindful of my presence. Nene was the one who seemed to be quite shy with me. While they ate, they talked about school events, their grades and the activities in which they had participated. The conversation was interrupted by the funny gestures of Iyat, the second to last child who most often brought laughter within the family. After lunch, Nene washed the dishes and took a short nap with her family. At 1.00 p.m., Tio Gusto left for work, and Nene for school. The younger children rested until 3.30 p.m. and then went out to play. Tia Lilia also went out to the balcony and breastfed her baby while she watched her children play in the yard.

The range of responses provided in answer to the question on leisure offers more interesting insights on the time allocation of husbands and wives. Of the seventy-six wives identified, 74 per cent indicated that they do take time for leisure. Among the husbands, a lesser number (49 per cent or 64.4 per cent) indicated having time for leisure. Interestingly, what a few women consider as leisure is actually economically profitable – *panahi* (sewing) for example, and *hilamon/pananom bulak* (weeding/gardening). The social-mindedness of the wives may be inferred from the other forms of leisure they identified: *lagaw-lagaw sa palibot-pamilya, attend sosyal o miting, estorya-estorya, bulig sa feeding* (roaming around the neighbourhood or visiting relatives, attending socials or meetings, helping in feeding). What the husbands consider as their leisure includes such diverse activities as engaging in or watching cock-fights, drinking, socializing and even cleaning their privately-owned *talamnan* (vegetable, rice or corn plots).

Cost Control and Financial Management

The relative flexibility of the sugar-cane workers may also be gauged by their cost minimization measures to enhance their chances of survival. In this connection, the survey findings show that the bulk of the respondent households' expenditure is for food (as much as 79.7 per cent).

In relation to the condition of workers on the different types of farm – that is, traditional, abandoned or transitional – the survey finding shows that the workers in the abandoned cane farms consistently spent less on food items as compared with those in traditional or transitional farms.

However, expenditure on alcohol was found to be highest among workers on the abandoned farms – a possible indication of a growing escapism in the form of alcoholism among unemployed workers. In the 1986 SRC study, Bañas noted that 7 per cent of the daily expenditures of the 'stayer' workers went to alcoholic beverages – a finding he considered significant, since the native drink (*tuba*) can be secured free, by the workers. Bañas (1986: 104) also found that some stayers take tuba whenever food is unavailable at their dinner table, a finding indicative of their attempt to numb away miseries. This finding that a comparatively great amount of the workers' income goes to food, confirms a Centre for Research and Communication report that between 60 per cent and 80 per cent of the income of rural households goes to food. While the CRC study estimated the average monthly budget of a family of five in Iloilo to be 1,492 pesos, the findings of the present study place the monthly food expenditure of the respondents at 823.2 pesos per month. A further analysis of the expenditure pattern shows the pattern of income deficit for workers on abandoned farms as well as for workers on farms administered by benevolent planters. As is to be expected, those on abandoned farms have greater deficits (as much as 31 per cent of their expenditure) than those on traditionally-managed farms. Those on transitional farms i.e. those released from full-time sugar-cane work, and engaged in their own agricultural production in lands they fully control, did not experience an income deficit. This is seen as proof of the value of the workers' gaining control of the means of production.

A further study of the daily expenditure of the respondent households reveals that 60.2 per cent of their food expenditure is for rice. Expenses for meat and vegetables constitute 34 per cent, while that for other food items, like sugar, is almost negligible (1.6 per cent). While only a minority of the households reported expenditure for snack foods, it is to be noted that those on the abandoned farms registered a mean of only 42 centavos (actually 0 median).

Comparative data on the pattern of food expenditure before the worsening of the crisis and at the time of the SRC-NEDA survey in 1985/6 were based on the memory recall of the respondents. Though subject to some flaws (such as faulty memory recall in some instances), the study revealed some interesting comparative data. For instance, the average weekly purchases of dried fish[1] in 1986 amounted to 49.5 pieces as compared to the average of 63.2 in 1983 (Bañas 1986: 106). The study further reveals that for all three groups of workers surveyed i.e. stayers, migrants and millworkers, the average volume of purchases decreased during the three-year period by as little as 26 per cent in the case of the migrants, and as much as 31 per cent in the case of the stayers and mill workers.

Both the SRC-NEDA and the VRC (Visayas Research Consortium) studies captured data on items annually purchased by sugar-cane workers. Bañas notes that from an average of about five pieces of clothes purchased in 1985, garment buying was reduced to about only three pieces in 1985/6. The subsequent upward trend in clothes purchasing may be inferred from the finding that in 1987 the average yearly purchase of clothes was 4.8 pieces. A comparable pattern may be observed from the reported purchase of slippers. As reported by Bañas, the volume of slippers purchased during the two periods (1983/4 and 1985/6), decreased from 28.3 pairs to 11.9 pairs, with the displaced millworkers reporting the greatest decrease of 41 per cent. During the milling season covered by the present study i.e. 1986/7, the average number of pairs of slippers bought was 8.5.

Pawning, Selling and Borrowing

The other adjustment mechanisms observed among sugar-cane workers in the 1985/6 survey include such activities as selling and pawning of household property. In the SRC-NEDA survey, 17 per cent of the migrants, and none of the stayers were reported to have engaged in this adjustment strategy after the last quarter of 1983. It was noted that a small proportion of the farm workers pawn or sell their personal or household effects because they are virtually without any other assets (Bañas 1986: 110). The situation for the millworkers is entirely different because the comparatively higher income which they received when the mill was in operation had enabled them to buy some moveable property. Yet, hardpressed for immediate cash, the displaced sugar-cane millworkers have been forced, by necessity, to sell or pawn whatever furniture and moveable property they had acquired. Among the items commonly sold or pawned are wristwatches, chinaware, electrical appliances, blankets, pillows, mosquito nets, and even some clothes.

Exchange of Goods, Resources and Services

One other indicator of the cane workers' flexibility which was gauged in the 1987 survey, is their social network system — that is, the web of social relationships which they can activate in times of need. Based on the survey finding, it appears that a functioning social network among the workers is based on the proximity of their social relations. In response to the open-ended questions posed regarding this network system, two surfaced – *kasilingan* and *kaingod*.

A *kasilingan* was defined by 33.3 per cent as somebody with whom they are on good terms; 25 per cent said this is someone who provides immediate assistance; 20.8 per cent regard a *kasilingan* as a person willing to help at all times; 18.1 per cent viewed the *kasilingan* as someone easy to approach. Thus, a *kasilingan* is obviously someone with social proximity. The workers made a clear distinction between a *kasilingan* and a *kaingod*, literally one's next door neighbour.

The data reveal that most of the workers who engage in reciprocal sharing of goods are relatives. Based on the survey findings, food ranks as the most frequently shared item among *kasilingan*. Shared a good deal less often are work or farm tools. Interestingly, the sharing of personal services came a poor third, perhaps indicative of the fact that they are assigned greater value by the respondents. In fact, a common observation made among the sugar-cane worker communities is their unwillingness to work on ventures which require the sharing of labour. The nature of sharing among the sugar-cane workers is delineated in the field notes of Da-anoy (1986a).

The field researcher noted that under extreme financial stress, even relatives cease to share their scarce resources willingly. Blood relatives who continue to do so often end up in conflict with their spouses.

When asked what forms of assistance they commonly solicit from their *kasi-lingan*, 44.5 per cent cited personal services or cooperation and sympathy. This form of help is commonly sought in times of emergency, such as when a member of the family has to be rushed to the hospital or when there is death in the household. It is to be noted that in the abandoned farm sites, the most common form of help sought is the sharing of food; less frequently, the lending of money. Apart from the type of assistance sought, the respondents were also asked to identify things which they borrowed from their neighbours. The survey revealed that the items commonly borrowed are food, work, tools, utensils, and money. Of the three groups of worker-respondents, those from the abandoned farms were most consistent with their answers. For these workers on abandoned farms, food and money rank as the two most commonly sought items from their neighbours. Borrowing is not as frequent as sharing among exchange partners. On an average, the respondents borrow from each other only once a week. Perhaps owing to their greater need, those in the abandoned farms borrowed more frequently i.e. twice per week.

Apart from inter-household sharing, there is also the prevailing practice of intra-household sharing – i.e. sharing among the members of the same household; the most commonly shared items being personal effects like everyday clothes, clothes for special occasions and shoes.

Reliance on Local Credit

Though some of the respondents borrow money from their neighbours, such loans are usually non-interest bearing, hence the repayment is usually equal to the amount of the loan. But since the loan acquired in this manner is often not enough to meet emergency needs, interest-bearing loans are sometimes also secured by the respondents. In fact, almost all of the respondents in the 1987 survey had approached another person or institution for this purpose. As reported, only five out of seventy-six respondents indicated that they did not borrow money from others. For those who acknowledged doing so, about a quarter reported turning to their *amo* (landlord; also employer of domestic helpers) for cash loans. As is to be expected, this pattern is more pronounced among workers in the traditional farms. If dependence on the landlord seems to be negligible in the transitional farm, it may be because the residents there are beneficiaries of the land transfer scheme of the former landowner (Lopez-Gonzaga 1985). Thus, whenever the need for money arises, the residents of the transitional farm approach either *kasilingan*, their children, or their association. However, the formal break in the relationship between the landlord and the workers may have paved the way for the entry of the *saka-an* (moneylenders) who charge a monthly interest of 20 per cent. This may be inferred from the fact that it is only among the respondents of Rosari-Nato, the transitional farm, where the *saka-an*, though subscribed to by only a minority, was reported as being one of the sources of credit.

When unable to secure credit from any of the aforementioned sources, the majority of the respondents (54.2 per cent) borrowed from neighbourhood stores, while a third borrowed from their relatives or their neighbours. For workers in the abandoned farms, however, the *kasilingan* ranks as the most popular alternative source of credit, followed by relatives and the neighbourhood store.

What is most revealing in the present study, is the finding that only 13.5 per cent of the respondents managed to avail themselves of the services of the formal banking institutions. This clearly indicates the inaccessibility of the financial institutions to these people; not withstanding the fact that the rural banks were established with the primary purpose of answering the banking needs of agrarian based communities. For those who have access to the banking facilities, 42.9 per cent make use of them to buy draught animals, while 28.6 per cent use the money for their children's education. Only a small percentage of the bank borrowers used their loans for less urgent concerns like the graduation expenses of a household member.

If only a few of the respondents were able to borrow from the banks, it is because only a tiny minority possess some moveable assets, like a *carabao*, to serve as collateral for prospective loans. Unsecured loans appear to be a rarity for the respondents, as may be inferred from the fact that a loan from the Social Security System (the SSS – quasi-banking institution) was extended to only one of them. Also, not one of the respondents had managed to secure a guaranteed loan from their landlord.

If the availability of collateral enabled few of the respondents to borrow from the banks, being bereft of collateral prevented others from having access to this credit resource, as pointed out by 50.8 per cent of them. A smaller proportion (28.7 per cent) refuse to borrow from the banks for fear that they may not be able to repay the loan. A few others (14.9 per cent) explained that they prefer borrowing from personal contacts because the amounts they need are not substantial. Though the different reasons given by the respondents are valid, they still do not negate the fact that the banking institutions, particularly the rural banks (because of their failure to provide a 'contextualized' lending programme) have generally failed to fulfil their mandate to serve the interests of the rural poor. As the survey findings show, the standard requirement of a loan collateral automatically disqualifies the majority of them from access to the banks' services. Also, the rural banks are usually located in the town centres, making them physically inaccessible to interior and upland based workers and peasants.

Summary and Conclusion

The excessive stress to which a society is subjected requires, on the part of its members (especially those passing through a crisis) a relative flexibility and adaptability in order for them to survive and, consequently, find an angle for recovery. The study delineates a number of relevant variables and indicators of the sugar-cane workers' flexibility (or inflexibility) and adaptability to crisis. In discussing the main theme of the study, comparative survey and participant observation data from previous research was presented to provide a time-frame and longitudinal view of the trend in the crisis situation. An integrative analysis of relevant findings of the SRC studies since the milling season 1982/3 has shown a consistent pattern of poorly managed sugar-cane workers' households, with low levels of educational attainment, low earnings, and large family size. Given these household characteristics, the workers have generally found it difficult to shape their destiny. For displaced workers who are not organized or linked to a strategic social grouping like a human development foundation or a workers' union, their lot has been locked. Extended case studies of households show that under crisis the workers' flexibility and adaptability are determined by other factors as well; like the number of working household members, the dependency burden, skills of members who are of working age, and the social network system. It was found that the absence of social connections deprives the unorganized workers of access to assistance programmes and even basic social services. Based on the survey findings on the previously-mentioned factors which determine household flexibility, it may be concluded that the sugar-cane workers of Negros Occidental, particularly those who are not organized and who live on abandoned farms, are grossly limited in their ability to cope effectively with the crisis.

As the sugar industry faced dissolution, what ceased to be peripheral, becoming instead an important factor in understanding crisis-coping, was the role played by women workers in the household. It was realized, from the study, that in times of crisis, the women of Negros contributed more than the men in ensuring the survival of their individual households. As the crisis deepened, they assumed sundry jobs and became petty entrepreneurs to meet the household's basic needs. How this initiative can, in turn, be translated as a catalyst for recovery bears some monitoring.

Despite this, women workers, particularly in the sugar industry, continue to face ingrained prejudice, and consequently are offered fewer employment opportunities on the sugar-cane farms. As a result of this continued stereotyping with regard to the work they can perform, they are still designated tasks mainly in weeding and fertilizing. Furthermore, women plantation workers are deprived of maternity leave benefits, and are not covered by Medicare.

Nevertheless, built into crisis phenomenon is the dynamic for decisive change, or a recovery angle. At the 'macro' level, a study of the key forces determining the trajectory of change in Negros Occidental shows some signs of a viable approach towards a recovery angle.

In spite of the polarization of different groupings, the coups of the past years have been followed by a period of relative restraint by the New People's Army and the vigilantes supported by right leaning groups. A perceptible shift from violence to moderation has been felt. Negros Province overwhelmingly supported the New Constitution proposed by the Aquino administration, and routed opposition candidates in the Congressional elections thereafter. These two developments are indicators of the continuing support of the political process by the Negrense. This promises change without violence.

Skies, in fact, seem to be bluer for Negros, with residents cautiously optimistic in the face of an apparent lull in the promised upswing in business activity. The ongoing milling season, the steady price of sugar and the prospect of good rice harvests are indicators of a new bullishness in the provincial economy. At the micro level, the possible angle of recovery is indicated in the household conditions of the sugar-cane workers, in the positive course of their income growth during the period from 1982 to 1983 through milling season 1986/7. Between 1984 and 1985 and 1985 and 1986, a drastic drop of 51.9 per cent in the monthly household income pegged against current and constant prices, was captured in the 1987 survey. When deflated by the regional Consumer Price Index, the percentage difference in wages in real terms fell by as much as 36 per cent. Also crucial at this level is the effect of the increased impact of women workers on household behaviour and, ultimately, on income patterns related not only to the household, but possibly to industry as well.

The 'Negros Day of Hope' held on 9 November 1988 captured a renewed faith and confidence in the future of the province. Viewed as sparks for such a renewal are the forecast of an increased sugar-cane production, the upward trend of domestic sugar prices, and the initial success of diversification efforts led by the entrepreneurial governor, Daniel Lacso, Jr. However, chances for the development of two much neglected sectors, namely women and the poor, will not increase without a change in the existing power structures and a relocation of control of the key resources, particularly land. Unless women and the poor are introduced into active participation, as vital elements of a production personpower, the economic growth of Negros will indeed be nominal, if not almost nil.

Note

1. Apart from their rice purchases, the respondents were also asked about their weekly purchases of fresh and dried fish. Since fresh fish are sold in various forms of measurement – i.e. by piece, by atado (bunch) or by kilo, a conversion rate of one third or thirteen pieces for each atado was used.

9

Women Plantation Workers and Economic Crisis in Cameroon

Piet Konings

Managements of tea estates have often given preference to female labour over male labour, on the assumption that women were *naturally* more suited to plucking tea (they had 'nimble fingers'). They were also thought to be cheaper and more docile than men. This managerial preference proved to be problematic, however, when tea farming came to Africa. First, it was difficult to recruit a regular and adequate supply of female labour for the newly created tea estates. The subordination of women under customary patriarchal controls in African societies often formed a serious obstacle to female migration and wage employment. Second, it was difficult to establish managerial control over women in the labour process. African women proved less submissive than management had expected. They demonstrated an ability to protect their interests both individually and collectively (Konings 1995b; Stichter and Parpart 1988).

In this chapter I focus on women pluckers at the Tole Tea Estate, one of the estates owned by the Cameroon Development Corporation (CDC).[1] The CDC is a large agro-industrial parastatal operating in the coastal area of Anglophone Cameroon. It was founded in 1946/7 for the purpose of developing and managing the approximately 100,000 hectares (ha) of estate lands which had been confiscated by the British Trusteeship Authority from German planters at the outbreak of the Second World War (Ardener, E. et al. 1960; Epale 1985; Konings 1993). Between 1946/7 and 1985/6, it virtually doubled its cultivated area from about 20,000 ha to 40,000 ha, with the aid of huge foreign loans. Today it is still the second largest employer in the country, surpassed only by the government itself. In the early 1950s it employed about 25,000 workers. At present, it has about 14,000 permanent workers and a few thousand seasonal and casual workers. Its chief crops are rubber, palm oil, bananas and tea. The CDC owns three tea estates and now enjoys a monopoly on tea production in the country.

The construction of the Tole Tea Estate in 1954 marked a turning point in the history of the CDC. This was its first estate to produce tea. More important for the present analysis, it was the first estate to recruit predominantly female labour. In

151

the first section below I examine what categories of women tended to sell their labour power to the estate management and how this relates to customary patriarchal controls in the local communities. In the second section I highlight the intensification of managerial control and exploitation of women pluckers during the present economic crisis. And in the final section I deal with the response of women pluckers to this severe crisis, showing that they have adopted various strategies to cope with the structural adjustment measures which have been planned and implemented by the management in close cooperation with the state-controlled trade union. What emerges from this study is that even during the economic crisis the management has failed to fully control the women pluckers in the labour process.

Women Tea Pluckers at the Tole Tea Estate

Upon the completion of the Tole Tea Estate, the management decided to employ women for plucking tea, the central activity in tea production. There were various reasons for this decision. Tea plucking had widely come to be identified as 'women's work'. Vast numbers of women had been recruited on tea estates in Asia and elsewhere on the assumption that female pluckers were more productive, more docile and cheaper than male pluckers. If women in Asia were plucking tea, why wouldn't women in Cameroon? Moreover, at the time the Tole Estate was opened, some women were already working on the CDC estates, mainly on a casual or seasonal basis. They were mostly the wives of estate workers, or women from the surrounding villages. Finally, the management was confronted with a serious shortage of male labour on the plantations during the fifties, due to the spread of non-estate coffee and cocoa production locally.

For a long time management efforts to recruit an adequate supply of permanent women pluckers proved disappointing. The recruitment drive formed a direct threat to customary patriarchal control over female labour. In 1952, shortly before the Tole Estate was constructed, Phyllis Kaberry published her classic study of women in the Bamenda Grassfields, the 'traditional' labour reserve for the coastal plantations. Throughout her book she highlights the contradictory position of women in society. While on the one hand, there is a general recognition that women play an indispensable role as child bearers and food producers, on the other hand, they are subordinated to patriarchal controls. These contradictions in women's position may not be as puzzling as they appear at first sight. Control over women's vital productive and reproductive labour constituted the basis of men's prestige, power and wealth in society. In a recent study of women in the Bamenda Grassfields, Goheen (1993: 250) observes:

Women grew the food crops and were expected to provide the necessities of daily life from their farms. Women's productive labour freed men to participate in (lucrative) trading networks; their reproductive labour increased the size of the household and thus the status and the labour force of the male head. Any surplus value women produced over and above that required for household needs and petty barter was in the hands of men, who retained all the profits.

Male elders and chiefs therefore had a vested interest in keeping women's productive and reproductive labour confined to their local communities, and they were inclined to resist female migration and wage employment.

Gradually, however, this patriarchal opposition could not forestall an increasing flow of female labour to the new estate: in 1988 the estate was employing 1,604 permanent workers, 63 per cent of them women. Some female workers are women who accompanied their husbands to the estate, but most are women without husbands, having escaped for one reason or another from the control of male elders in their local communities and eager to build up an autonomous existence (see Table 9.1). Some of these are young, single women who had no access to land, but were expected to work as unpaid labour on their family farms. The majority are older women who, following the dissolution of a marriage either by death or divorce, lost relatively secure usufruct rights. Rather than to become dependent on the family elders for their survival, they preferred to migrate to the estate, where they could be sure of a regular monthly wage income.

Women who seek employment on the estate are by and large illiterate, or at least poorly educated. Plantation labour is one of the rare employment opportunities in the capitalist sector for this category of women. Female tea estate workers therefore have a high stake in plantation labour, especially since most of them are not only 'husbandless' (Bryceson 1980), but also household heads with children to support. They are deeply dependent on their income from plantation work for the reproduction of their families. As a result, they tend to be highly committed to their jobs, becoming increasingly stabilized (cf. Safa 1979: 447–8). Over 60 per cent of the Tole women have been employed on the estate for more than ten years. This workforce stability is enhanced by the fact that most women maintain few close links with their region of origin. The 'husbandless' women in particular seem to have pretty much severed the ties with their home towns and villages. Their escape from patriarchal control generally prevents them from maintaining contact with their family members and gaining access to land at home. Their 'exit-option' from wage employment would be for the most part a switch to petty trade in the coastal urban centres, and not a return home. Even married female workers on the estate find it hard to keep up intensive contacts with their family members at home. Geographical distance and a six-day working week pose serious obstacles to regular communication. Many female workers visit their home towns only during the annual leave period, or on special occasions such as funerals of close kin.

Table 9.1. Demographic Characteristics of Tole Women in 1988. Per Cent.

Age	
15–25 years	7.5
25–35 years	32.5
35–45 years	41.0
45–55 years	17.0
55 years and older	2.0
	100.0
Marital Status	
single	9.0
married	32.0
widowed	27.0
divorced/separated	25.0
free union	7.0
	100.0
Educational Level	
illiterate	78.0
primary education	20.5
post-primary education	1.5
	100.0
Religion	
Christian	94.0
ancestral belief	6.0
	100.0
Years of Service	
0–5 years	17.0
5–10 years	19.5
10–15 years	39.0
15–25 years	19.5
25 years and above	5.0
	100.0

Source: Computed from data supplied by the estate office.

The female labour force on the estate by no means forms a homogeneous group. One can observe distinct divisions within it, based mainly on education, occupation, marriage, or ethnic and regional origin. There is a division, for instance, between the vast majority of illiterate and poorly educated field workers and the tiny group of better educated clerical workers. The latter tend to look down on the field workers and treat them rudely when they visit the estate office or clinic. There is also a division between married and unmarried female workers. Unmarried women are relatively free from customary patriarchal controls, but they are still held in low regard, being labelled as 'loose women' or 'prostitutes'. Married female workers are more respected, yet they are subjected to considerable male controls. Some husbands insist that according to African tradition, women are responsible for the upkeep of the family; these men are likely to make very little financial contribution, spending their wages as they choose. Others, in particular men from the Bamenda Grassfields, go as far as to claim that they are traditionally entitled to their wives' wage income. One more division is on ethnic and regional lines. On the estate there is a clear divide between women from the Bamenda Grassfields and women from the coastal/forest region: 67 per cent of the women originate from various ethnic groups in the Bamenda Grassfields and 27 per cent from ethnic groups in the coastal/forest region. Although ethnic and regional heterogeneity forms a potential source of conflict, with rivalry being sparked by suspicions of favouritism in hiring and promotion or in disagreements with supervisors or co-workers, there has been little incidence of serious, extended ethnic clashes. Generally speaking, estate workers of different ethnic groups or regions live and work together peacefully. It has been the consistent policy of the management, as well as church and union leaders, to mobilize and organize workers on a multi-ethnic basis, and this policy seems to have created a certain measure of understanding and tolerance among the workers for each other's sociocultural background, and to have fostered bonds of companionship and friendship across ethnic boundaries. The general use of Pidgin English has also helped to overcome communication barriers between the various ethnic groups. The most important reason for this relative harmony, however, appears to be the sharing of similar living and working conditions on the estate, which is a classic example of an occupational community.

Finally, there is still another division between male and female workers on the estate. The root of this cleft is the pervasive ideology of male dominance. For example, the men expect the male-dominated management to give them preference over women when there is an opportunity for further training or promotion, and to fire female workers when there is a retrenchment exercise. They are reluctant to work under female supervision, and they regularly harass their female colleagues. There is also a persistent fear among the married men that wage employment will make their wives too 'headstrong' or independent.

Remarkably, these divisions within the Tole labour force have never kept the workers from engaging in common actions for the promotion of their interests. They display an ambivalent attitude towards plantation labour: they value it, since it provides them with a regular source of income; but they harbour an acute feeling of exploitation and subordination in the workplace.

Intensification of Managerial Exploitation and Control during the Economic Crisis

As in other parts of the world (Loewenson 1992), estate workers in Cameroon receive low remuneration for their arduous work. What is more, female workers were initially paid less than male workers. Apparently the management wanted to implement the same system of unequal payment for men and women that prevailed on many other tea estates in Asia and Africa (Kurian 1982; Vaughan and Chipande 1986). This was justified on the spurious grounds that female workers needed less time (seven hours) to complete their daily task than did male workers (eight hours). In 1967, however, a Federal Labour Code was promulgated which prohibited employers from paying men and women differently for equal work, and the estate management then introduced equal rates of pay.

Prior to the economic recession, tea pluckers were earning an average net monthly wage of approximately FCFA 20,000–25,000.[2] Some pluckers were unable to earn this much, however, mainly due to a linkage between the remuneration of workers and the system of task work prevailing on the plantations. Completion of the daily task set by the management entitled a worker to the daily basic wage; non-completion was punished by pro-rata payment, a proportion of the wage equivalent to the proportion of the task completed.

The management has never denied that women receive low cash wages, but it continually stressed that the variety of non-cash benefits supplied by the corporation, such as accommodation, medical care, and a plot for food cultivation, supplement their incomes. Some remarks are in order here, however. The dwellings provided are very small and are in a deplorable state. The clinic lacks qualified personnel and suffers regular shortages of essential drugs. During my fieldwork I found that only 57 per cent of the female workers had been allocated land free of charge for food production. The remaining women had to hire land from local peasants.

The exploitation of women workers has intensified during the economic crisis that has been affecting the corporation since 1986/7. Between 1986/7 and 1990/1, the CDC suffered losses totalling about FCFA 19 billion. There is no doubt that the sharp fall in commodity prices on the world market was primarily responsible for the virtual bankruptcy of the corporation. Nevertheless, other factors also contributed to the emergence and persistence of the crisis. First, there was the political elite's inability or unwillingness to control the imports of cheap tea, which caused CDC sales to stagnate in the domestic market. Second, there have been frequent reports of the managerial elite's involvement in massive embezzlement, reckless expenditure, waste and power struggles; small wonder that many estate workers did not believe the management would be capable of effectively combating the crisis (Konings 1995a).

To save the company from total collapse, the CDC management and the state-controlled trade union agreed to adopt a series of adjustment measures aimed at cost reduction and productivity increases. On 23 August 1987 they agreed upon a substantial intensification of task work. The daily quota required from tea pluckers was upped from 26 to 32 kg of green leaves. This was accompanied by a redoubled managerial crusade against 'undisciplined and unproductive' workers:

> The corporation is required to meet certain standards of efficiency and to be self-supporting and profitable . . . That is why we have to be generally very strict on discipline and sanction any manifestation of laxity . . . Maximum efforts shall be required of employees so as to continue producing more at lower and lower cost; laxity and *laissez faire* which are characterized by an alarming rate of absenteeism and uncompleted tasks shall not be tolerated. Pilferage shall meet with maximum sanctions.[3]

As the corporation's financial position continued deteriorating, the management proposed further 'austerity' measures to the union leadership. Following negotiations a new agreement was signed on 6 January 1990 entailing drastic cuts in the wages and fringe benefits of all workers, amounting to some 30 to 40 per cent of their previous incomes. Henceforth the workers would be obliged to make substantial contributions towards the range of services which the corporation previously supplied free of charge, such as housing, water, electricity and medical facilities. The most draconian measure was the introduction of a compulsory savings scheme, forcing the workers to save at least 15 per cent of their basic wages and salaries to aid the corporation's recovery.

The union leadership had expected the workers' increased output and financial sacrifices to forestall, or at least minimize, any retrenchment. This proved wishful thinking. The management soon embarked on mass layoffs. Between 1987/8 and 1990/1, the labour force on the Tole Estate shrank from 1,604 to 974.

As a consequence of the various adjustment measures, women's position on the estate has deteriorated dramatically. They are compelled to work harder for a much lower reward. They experience immense difficulties in combining plantation work with their other productive and reproductive responsibilities. Owing to their increased workload on the estate, they must stay longer in the field. They now often lack time and energy to engage in additional income-generating activities, such as food production and trade, and to care for their children. The estate does provide a crèche and a nursery school, at a cost of FCFA 2,000 a month, but very few children attend them. Most women are dissatisfied with the services rendered there, and they prefer to use other methods of child care – relying, if possible, on their older children, relatives, friends or neighbours. The period of breast-feeding is the most problematic. The management does not allow the women to take their nursing children to the field, though a few women do it all the same. Usually women go home during breaks (if they are not too far away) to tend to their children or give them a quick feeding. It is not uncommon, however, to see other children bringing hungry babies to their resting mothers during breaks (DeLancey 1981). Last but not least, the women's employment on the estate has become insecure, due to the repeated reorganization measures.

Response of Women Pluckers to the Economic Crisis

The Tole women challenge the pervasive managerial assumption that female workers are more docile than males (Elson and Pearson 1984). From the outset the management had trouble controlling the women pluckers (Konings 1995b), and the Tole Estate quickly acquired a bad reputation in management circles. Time and again the CDC management was forced to admit that 'the Tole Tea Estate is one of the most troubled spots on the CDC plantations'[4] and that 'output on the estate has remained almost pathetically low'.[5]

Management's apparent lack of control over women pluckers does not come as a complete surprise, for African women have persistently shown a capacity to protect their interests individually and collectively. Many accounts have revealed frequent conflicts between husbands and wives over land, labour and 'capital' (cf. Kaberry 1952). African women have regularly engaged in informal and collective modes of resistance to male abuse of power (Isaacman 1990). Individual informal actions have included refusal to cook or have sexual intercourse, withdrawal of domestic and agricultural labour, temporary or permanent return to the parental home, spreading gossip about their spouses, migration, calling in the supernatural, and manipulation of men. Collective actions of women have occurred in several African countries, including Cameroon (Ardener 1975; O'Barr 1984; Parpart 1988). Two such actions, known as *anlu* and *fombuen,* have been documented in the Bamenda Grassfields, the region of origin of most Tole Tea women (Nkwi 1985; Diduk 1989; Shanklin 1990). In essence these were manifestations of women's resistance against male abuse of power and against the colonial power's efforts to establish control over the means of production (land) and the labour process, thus forming a direct threat to women's food production and relatively autonomous work activities. Tole female pluckers have likewise proved capable of undertaking a whole range of actions against their subjection and exploitation in the labour process, including syndicalist, collective, and informal actions (Cohen 1980; Stichter 1985).

Similarly to other parts of the world (Elson and Pearson 1984; Pittin 1984; Ecevit 1991), women's involvement in trade union affairs at the Tole Estate has usually been low. They participate little in the decision-making process and rarely compete for office. This must be attributed mainly to their illiteracy, their lack of time due to multiple productive and reproductive responsibilities, and the union's patriarchal structure. Any support for the union has always been highly conditional and instrumental. If the union leadership and shop stewards act promptly and forcefully upon receipt of a complaint and 'deliver the goods', they can be sure of the women's support. Tole women have always admired militant trade union leadership. They massively joined the 1955 and 1966 general strikes of CDC workers called out by the union for an improvement in wages and conditions of service. Conversely, if the union leadership fails to represent the interests of the workers in general and the women in particular, Tole women have been inclined to circumvent the union. In particular, the establishment of state control over the union in 1972 touched off a crisis of confidence in the union leadership and shop stewards.

There is ample evidence, moreover, that the Tole women have sought alternative ways of defending their interests, notably by engaging in informal and collective actions, when they perceive that the union is unresponsive to their demands (Konings 1995b). Informal modes of labour resistance, such as insubordination, output restriction, unauthorized absence and theft, are obviously less visible and dramatic forms of protest than collective modes of resistance, but they occur almost daily on the estate. Collective actions such as strikes, riots, demonstrations and go-slows, are riskier in Cameroon, where the post-colonial state strives for 'total' control over civil society (Bayart 1979) and has virtually outlawed strikes since 1967. Strikes occur only as a last resort, when institutionalized procedures for settling conflicts between management and labour, such as joint negotiation and collective bargaining, are either absent or ineffective.

Interestingly, despite increasing state and management controls over the labour process, Tole women have continued to display a remarkable degree of militancy. They have engaged in a considerable number of informal and collective actions either to back up demands for improvement in wages, fringe benefits and incentive bonus schemes or to protest against increased control and exploitation. They have been most successful in resisting managerial efforts to raise the level of task work: plucking norms were hardly increased by the management in the period 1963–80. They have been least successful at removing authoritarian or overzealous managerial and supervisory staff from their posts. Their actions should not be judged only in terms of success or failure, of course. They should also be conceived as key opportunities for raising women's capacities for self-organization and self-confidence.

Given this long tradition of women's militancy on the estate, what was their reaction towards the severe economic crisis and adjustment measures?

Confronted with the close cooperation between management and union leadership in planning and implementing austere adjustment measures, the response of the women pluckers has been complex and varied. Some women have opted for a single strategy, others for several strategies, either simultaneous or consecutive. A growing number of women have become survival-oriented in the climate of insecurity. They are inclined to acquiesce in whatever economic recovery measures management may introduce, for the sake of keeping their job. They try to impress the management with above-average output, and they avoid conflicts with their supervisors. This tendency is fuelled by the fact that most Tole women do not maintain close links with their home area, thus lacking an easy 'exit-option'.

Some women still rely on the bargaining strength of the state-controlled trade union for the protection of their interests. But most women have lost whatever confidence they ever had in the union leadership. They tend to bypass the union and lodge their individual or collective complaints with the Labour Department, which is usually perceived by workers as a neutral intermediary between management and labour. This department is overburdened with work, often resulting in long delays in settling disputes.

Unexpectedly, the majority of the women continue to resort to both informal and collective protest actions. These are directed against their intensified exploitation and subordination manifest in managerial efforts to raise labour productivity by intensifying task work, to strengthen control over the labour process and to minimize wages and other conditions of service. One can nevertheless observe a decline in the number of collective actions. This is understandable, for in times of economic crisis strikes are more likely to elicit severe managerial reprisals, in the form of summary dismissals, than in times of prosperity. Significantly, the decline of collective actions appears to have been compensated by an increase in the number and intensity of *informal* actions, which are difficult for state and management to control even during the economic crisis. For example, insubordination seems not only to have increased during the crisis, it is also more frequently accompanied by insults and assaults on supervisory staff.

Women Pluckers and the Maximization of Labour Productivity

In January 1987 the management and the union leadership agreed on an increase in task quotas from 26 to 32 kg of green leaves. While the new norms were implemented on the two other CDC tea estates in June–July 1987, women pluckers on the Tole Estate continued to resist this decision. Eventually the CDC General Manager personally called on the union leadership to persuade the Tole pluckers to accept the decision. It was not until October that the Tole pluckers reluctantly accepted the increase in task work at a mass meeting organized by the union.

One year later, on 3 October 1988, the Tole pluckers went on strike. They argued that they were unable to fulfil the new task requirement, even though they worked from 6.30 a.m. to 6.00 p.m. every day. The union president strongly condemned their strike action, saying that it was illegal because they had not followed the prescribed procedures for settling conflicts. He was booed and jeered by the angry women. The CDC General Manager then threatened the strikers with dismissal if they did not report for work the next day, with the result that most women decided to resume work. Seventy-nine continued striking, and they were summarily dismissed. Many of them stayed in the area afterwards, surviving on farming and trading.

No further collective actions on this issue have been reported. There is some evidence, however, that the pluckers increasingly resort to informal actions to protest against the managerial attempts to raise labour productivity. Estate reports suggest that since the onset of the crisis there has been a rise in various forms of informal resistance, such as late arrival at work, absenteeism, malingering and poor work performance. The rate of absenteeism in particular has caused concern in management circles. Women tend to be absent from work more often than men because of their multiple responsibilities outside the workplace. The estate manager reported to the union that absenteeism could be blamed for the loss of over 1,700 person-days during the months July–August 1992. He estimated the corporation's financial loss from this at about FCFA 11 million. He warned the union president that he would not hesitate to dismiss over a hundred workers if the union and shop stewards failed to address this situation.[6] Apparently the managerial crusade against 'undisciplined and unproductive' workers had not been successful during the crisis.

Women Pluckers and Intensified Managerial Control

Supervisory staff constitute an essential link between the estate management and the workers. They are expected to communicate management orders to the workers and to maintain discipline in the workplace. While the command structure on the estate has remained male-dominated, one may observe a change in the gender composition of the supervisory staff in the female-dominated plucking section. The position of overseer was formerly monopolized by men, but the position of head of the separate labour gangs was open to women at an early stage. Gradually some women were appointed overseer and even senior overseer. Despite this change, problems of control have persisted.

Since the onset of the economic crisis there has been enormous management pressure on the supervisory staff to exercise tight control over their subordinates, to use their authority to the full and, if necessary, to discipline any worker without compassion. The increasingly authoritarian and overzealous supervision appears not to have had the desired effect of strengthening managerial control at the workplace. It seems instead to have antagonized the women, making them feel like slaves, incessantly prodded to increase production without reward. Although there used to be numerous collective actions aimed at the transfer or dismissal of unpopular supervisors, such actions abated during the economic crisis. The women seem to have realized that such actions not only are risky, but they will almost certainly be ineffective. With management actively encouraging strict supervision during the crisis, it is not likely to give in to collective demands from the workers for supervisors' removal.

However, there does seem to be widespread informal resistance against intensified managerial control. It manifests itself in various forms, including:

Insubordination

It is not uncommon for women to simply refuse to carry out orders from their supervisors, even if such a refusal might result in considerable loss of output. For instance, some women refuse to do overtime on Saturdays during the peak plucking season, as they see Saturday afternoons as a period for working on the food farms, trading and washing clothes. Estate records suggest that insubordination is increasingly accompanied by insults and abuse, and even physical attacks on supervisory staff.

Attempts to discredit supervisory staff

At times, women try to play tricks on supervisors and to discredit them with their superiors. This often takes the form of falsely accusing them of bribery, corruption and other misdeeds, in the hope that management will take action against them.

The threat of using mystical powers

Women sometimes threaten to harm or kill a supervisor by sorcery and witchcraft. It is hard to know whether this is mere bluff or a real threat.

Women Pluckers and Minimization of Wages and Other Conditions of Service

Women pluckers have always fought for improvement in their living standards, and they have steadfastly resisted any managerial effort to curtail their wages and fringe benefits. It is little wonder, then, that they were enraged about the 6 January 1990 agreement between management and union leadership which introduced a compulsory savings scheme and brought about a drastic reduction in their fringe benefits. Their bitterness was reflected in a collective action two years later.

In early 1992, Tole women pluckers, together with workers on other CDC estates, began agitating for the abrogation of this agreement and the immediate payment of seven quarters' family allowance arrears. When the management tried to employ delaying tactics, they went on strike from 21 to 26 May 1992. As a result of this strike, the management agreed to various amendments to the agreement, including the re-introduction of free accommodation and water, as well as a 24 per cent reduction in medical costs.

In addition to this collective action, women pluckers have engaged in many kinds of informal actions in protest against low remuneration. These have included output restriction, sabotage and involvement in 'illegal' income-generating activities. Some women do not keep to the plucking standards. They add bad leaves to the good ones, a practice which enables them to complete their task earlier and to achieve more weight and income. Others steal tea from the factory and sell it to middlemen. The management frequently complains that the theft of tea has reached unprecedented levels since the economic crisis and has caused serious losses to the company. Still other women cut tea bushes and pruning to use them for firewood. These and other 'illegal' activities are not only expressions of protest, but also they provide welcome supplements to the women's meagre incomes.

Conclusion

In this study I have attempted to show that plantation work plays an ambivalent role in the lives of women tea pluckers in Cameroon. It has positive meanings for them in that it enables women, especially those who are illiterate and 'husbandless', to escape patriarchal controls in their local communities and achieve a relatively autonomous existence. As household heads, they are highly dependent on plantation work for the reproduction of their families, and they display a high commitment to their job. But the work also has negative consequences. It is very tedious and poorly paid, and it is unlikely to foster much job satisfaction. It requires so much energy and time that the women have difficulty coping with their other roles outside the workplace. Above all, it fails to provide them with adequate income, and it offers few prospects for promotion and social advancement. These negative features are little conducive to labour productivity, and they pose problems of labour control.

Tole women have never acquiesced in control and exploitation in the workplace. They have engaged in various modes of resistance, thus challenging the common managerial assumption that female workers in developing countries are easier to control than male workers. Women plantation workers in other African countries have demonstrated a similar capacity to defend their interests against employers. Presley (1986) reports that Kikuyu women employed on settler farms in Kenya were involved in labour protests from the 1920s to the 1960s. With a persistent series of work stoppages and strikes, mostly aimed at improving wages and working conditions, these women gained confidence in the power of their collective efforts. This power was reinforced by the heightened political militancy of post-war Kenya. Research on the agro-industrial plantations of Bud in Senegal (Kane 1977; Mackintosh 1989) and in Nigeria (Jackson 1978) has highlighted women's involvement in various forms of action. Class-oriented struggles of women plant-ation workers have also been reported in Tanzania (Mbilinyi 1988).

This study provides evidence that the majority of the Tole female pluckers have not abandoned their 'traditional' militancy even during the economic crisis, and that they continue to engage in (sporadic) collective actions and (frequent) informal actions to protest against their intensified exploitation and subordination in the workplace. A number of factors appear to account for their ongoing resistance. One of these is their remarkable solidarity. The various internal divisions among the women never seem to have impeded them from undertaking common actions against the employer, though it cannot be denied that their solidarity has been somewhat weakened by the present economic crisis. This solidarity has been fostered by their residing in an 'occupational community' where they share similar living and working conditions, speak one lingua franca (Pidgin English), and can easily communicate with each other. These poorly educated women have always put less faith in the institutionalized bargaining procedures than the more moderate union leadership, which is composed mainly of male clerks and supervisory staff. The repeated failure of such procedures to achieve workers' aims, together with the establishment of state control over the unions, has strengthened this belief in militant action as the only way to bring management to its knees and redress long-standing grievances.

The growing demotivation of many women during the crisis has also played a role in the rising militancy. Far from being rewarded for their higher productivity by increased wages, women have seen their real incomes seriously eroded. Most women are still reluctant to resign and forfeit their regular monthly wage income, however meagre; but a growing number of women are no longer interested in keeping their job at any cost, especially since they have also lost confidence in the corporation's eventual economic recovery. Some have already resigned, collecting their long-service awards and gratuities as well as their voluntary and compulsory savings. They then invest this capital in farming, trading, and other potentially lucrative activities. Others are still contemplating resignation. No longer committed to their job, they tend to become either more militant and less reluctant to offer resistance, or more apathetic, having thought out their alternatives.

Another factor influencing women's attitude is their daily exposure to the glaring socioeconomic inequalities on the estate. Although the corporation's adjustment measures have effected a sizable reduction in the high salaries and numerous fringe benefits of the managerial staff, there remains an immense gap in incomes and living standards between them and the workers. There is no doubt that women's intense feelings of exploitation and subordination are reinforced by the blatant contrast between their own misery and managerial opulence.

The management has failed to fully control women's resistance during the crisis. There is no reason to believe it will succeed in doing so in the near future. Much to the contrary, there are indications that labour resistance may even be on the increase. Recently the unions regained most of their former autonomy, and the union leadership has grown more critical of the adjustment measures that it had previously agreed on with management. CDC workers have traditionally admired militant union leadership, and they will be inclined to support union actions in defence of their interests.

Notes

1. My research on women pluckers was part of a larger project on plantation labour in Anglophone Cameroon. Between 1985 and 1993 I regularly visited the Tole Tea Estate.
2. FCFA 100 is equal to French Franc 1.
3. See Report of the Consultative Meeting with the Ministry of Labour and Insurance at Provincial Level by Mr. P.M. Kamga, dated 28 October 1989, in File MEPS/SWP/ BU.134, Vol. IV, General Correspondence CDC.
4. Report of the Conciliatory Meeting between CDC Management and Tole tea workers on 25 November 1976, in File MEPS/SWP/BU.124/S.2, Industrial Trade Disputes-CDC.
5. See CDC, Tole Tea Estate, Minutes of Estate Consultative Meeting held at the Club on 13 October 1990.
6. See letter from Mr. R.M. Achiri, Estate Manager Tole Tea, to President Agricultural Workers' Union, dated 16 September 1992, in File MTPS/IDTPS/SWP/LB.2/Vol. XXVII, Complaints from CDC.

Bibliography

Ali, A. 1980. *Plantation to Politics – Studies on Fiji Indians*, University of the South Pacific: Suva

Andrews. C. F. 1918. *Indian Indentured Labourer in Fiji*, The Colurtype Press: Perth

ARAGI 1884 to 1921. *Annual Reports of the Agent General of Immigration in Fiji Royal Gazette, 1887–1921*, Suva

Ardener, E., Ardener, S., and Warmington, W.A. 1960. *Plantation and Village in the Cameroons*, Oxford University Press: London

Ardener, S. (ed.) 1975. *Perceiving Women*, Malaby Press: London

Bañas, R. 1986. The Living Condition and Adaptation Strategies of the Displaced Sugarcane Workers of Negros. Final Report Submitted to the National Economic and Development Authority, Region VI Illoilo City, La Salle – Social Research Center: Bacolod, pp. 63, 18, 90, 104, 106, 110 (*mimeographed*)

Bayart, J. F. 1979. *L'Etat au Cameroun*, Presses de la Fondation Nationale des Sciences Politiques: London

Beall, J. 1991. 'Women Under Indenture in Natal', in Surendra Bhana (ed.), *Essays on Indentured Indians in Natal*, Peepal Tree Press: Leeds, pp. 89–115

Beckford, G.L. 1972. *Persistent Poverty: Underdevelopment in Plantation*, Oxford University Press: New York

Beckford, W. 1773–84, *Plantation Papers of William Beckford, 1773–1784, C.* 107, 43. Public Records Office: London

Beckles, H. 1989. *Natural Rebels: A Social History of Enslaved Black Women in Barbados*, Rutger University Press: New Brunswick and New Jersey

Bennett, J.H. and Cary Heylar, 1964. 'Merchant and Planter of Seventeenth Century Jamaica', *William and Mary Quarterly* 3(21): 53–76

Benston, M. 1969. 'The Political Economy of Women's Liberation', *Monthly Review* 21(4): 13–27

Best, L. 1968. 'A Model of Pure Plantation Economy', Paper presented at a West Indies Agricultural Economic Conference Jamaica (*mimeographed*)

Bhowmik, S. not dated. 'A Sociological Study of Tea Plantation Labour in Eastern India: A Case Study in the Dooars of West Bengal', Ph.D. thesis, Delhi University

—— 1981. *Class Formation in Plantation System*, People's Publishing House: New Delhi

BM 1838. British Museum Additional Manuscripts 51819 and 27970. London

Brass, T. and Bernstein, H. 1992. 'Introduction: Proletarianisation and Deproletarianisation on the Colonial Plantation', in E. Valentine Daniel, H. Bernstein and T. Brass (eds), 'Plantations, Proletarians and Peasants in Colonial Asia', Special Issue of *The Journal of Peasant Studies* 19(3 and 4): 1–40

Breman, J. and Valentine Daniel, E. 1992. 'Conclusion: The Making of a Coolie', in E. Valentine Daniel, H. Bernstein and T. Brass (eds), 'Plantations, Proletarians and Peasants in Colonial Asia', Special Issue of *The Journal of Peasant Studies* 19(3 and 4): 268–95

Brereton, B. 1974. 'The Experience of Indentureship 1845–1917', in J. LaGuerre (ed.), *Calcutta to Caroni,* Longmans: Caribbean

—— 1979. *Race Relations in Colonial Trinidad 1870–1900,* Cambridge University Press: Cambridge

—— 1981. *A History of Modern Trinidad 1783–1962,* Heinemann: London

Bruce, C.A. 1839. 'Report on the Manufacture of Tea and on the Extent and Produce of Tea Plantations', *Assam. Journal of Asiatic Society of Bengal* 8: 497–526

Bryceson, D. 1980. 'The Proletarianization of Women in Tanzania', *Review of African Political Economy* 17: 4–27

Burawoy, M. 1980. 'Migrant Labour in South Africa and the United States', in T. Nichols (ed.), *Capital and Labour: A Marxist Primer,* Fontana Paperbacks: Glasgow

Bush, B. 1990. *Slave Women in Caribbean Society,* Indiana University Press: Bloomington, Indiana

C.G.F. Central Government File, Jamaica Archives. Files 1B/9/3- 1B/9/166 are relevant. The file referred to in the text is identified by its number

CLC 1935. *Ceylon Labour Commission Handbook,* Dodson Press: Colombo

Clarke, E. 1979. *My Mother Who Fathered Me,* George Allen and Unwin: London

COF or C.O.R. 1835–1917. Colonial Office Files or Record (referred in the text by relevant file nos.). Public Records Office: London

Cohen, R. 1980. 'Resistance and Hidden Forms of Consciousness amongst African Workers', *Review of African Political Economy* 19: 8–22

Comins, D.W.D. 1893. *A Note on Emigration from India to Trinidad,* Bangal Secretariat Press: Calcutta

Cooper, Thomas 1824. *Facts Illustrative of the Condition of the Negro Slaves in Jamaica,* London

Courtenay, P.P. 1965 and 1980. *Plantation Agriculture,* Bell: London (Revised edition in 1980, West-Minister Press: Colorado)

Craton, M. and Walvin James 1970, *A Jamaican Plantation: The History of Worthy Park, 1670–1970,* W.H. Allen: London and New York

CSO 1909 to 1916. *Colonial Secretary's Office Documents, C-Series and Minute Papers, 1909–1916,* pertaining to reports of the Agent General of Immigration in Fiji: Suva

CSP 1716–17. Colonial State Papers Collection. Public Records Office: London

Cumpston, I.M. 1953. *Indians Overseas in British Territories,* Oxford Historical Series, Oxford University: London

Da-Anoy, Mary Angeline 1986a. The Sugarcane Workers' Household Under Crisis. Field Notes #02-1986, (*Mimeographed*) La Salle – Social Research Centre: Bacolod

—— 1986b. Out-farm Employment. Field Notes #01:08-1986,(*Mimeographed*) La Salle – Social Research Centre: Bacolod

Dalla Costa, M. 1973. 'Women and the Subversion of the Community', in *The Power of Women and the Subversion of the Community,* Falling Wall Press: Bristol

Daniel, E. V. 1981. Tea Talk: Measures of Labour in the Discourse of Sri Lanka's Estate Tamils, (*Mimeographed*)

Daniel, E. V., Bernstein, H. and Brass, T. (eds) 1992. Plantations, Proletarians and Peasants in Colonial Asia, *The Journal of Peasant Studies* 19 (3,4)

de Silva, K.M. 1979. 'Resistance Movements in Nineteenth Century Sri Lanka', in M. Roberts (ed.), *Collective Identities, Nationalism and Protest in Modern Sri Lanka*, Marga Institute: Colombo

DeLancey, V.H. 1981. 'Wage Earner and Mother: Compatibility of Roles in a Cameroon Plantation,' in H. Ware (ed.), *Women, Education and Modernization of the Family in West Africa*, Australian National University: Canberra (Changing African Family Project Series, monograph no. 7, pp. 1–21)

DLS 1977. *Employment Survey*, Department of Labour: Colombo

Diduk, S. 1989. 'Women's Agricultural Production and Political Action in the Cameroon Grassfields', *Africa* 59 (3): 338–55

DPRO 1915–18. Documents of the Public Records Office, pertaining to Indian Immigration to the West Indies (referred to in the text by file no.), London

Dube, L. 1986. 'Introduction', in L. Dube, E. Leacock, and S. Ardener (eds), *Visibility and Power: Essays on Women in Society and Development*, Oxford University Press: New Delhi

Dube, L., Leacock, E. and Ardener, S. (eds) 1986. *Visibility and Power: Essays on Women in Society and Development*, Oxford University Press: New Delhi

Ecevit, Y. 1991. 'Shop Floor Control: The Ideological Construction of Turkish Women Factory Workers', in N. Redclift and M.T. Sinclair (eds), *Working Women: International Perspectives in Labour and Gender Ideology*, Routledge: London and New York, pp. 56–78

Edholm, F., Harris, O. and Young, K. 1977. 'Conceptualising Women', *Critique of Anthropology* 3 (9 & 10)

Edwards, B. 1801. *The History, Civil and Commercial of the British Colonies in the West Indies*. 3 Vols, J. Stockdale: London

Elson, D. and Pearson, R. 1984. '"Nimble Fingers Make Cheap Workers": An Analysis of Women's Employment in Third World Export Manufacturing', in P. Waterman (ed.), *For a New Labour Internationalism*, Ileri: The Hague, pp. 120–41

Emmer, P.C. 1985. 'The Great Escape: The Migration of Female Indentured Servants from British India to Surinam, 1873–1916', in D. Richardson (ed.), *Abolition and its Aftermath: The Historical Context, 1870–1916*, Frank Cass: London, pp. 245–66

Engels, F. 1978. The Origin of the Family Private Property and the State, in R.C. Tucker (ed.), *The Marx-Engels Reader*, Norton and Company: New York

Epale, S.J. 1985. *Plantations and Development in Western Cameroon, 1885–1975*, Vantage Press: New York.

FCR 1911–21. *Fiji Census Reports*, Government Printer: Suva

Ferguson 1866–1868. *Fergusons' Ceylon Handbook and Directory*, A. M. and J. Ferguson: Colombo

FLCP 1920. *Fiji Legislative Council Paper No. 46*, Fiji Royal Gazette: Suva

Fox-Genovese, E. 1982. 'Placing Women's History in History', *New Left Review* 133: 5–29

French, J. and Ford-Smith, H. 1986. Women's Work and Organizations in Jamaica, 1900–44. (*Mimeographed*)

Gardiner, J. 1975. 'Women's Domestic Labour', *New Left Review* 89

Gillion, K.L. 1962. *Fiji's Indian Migrants: A History to the End of Indenture in 1920*, Oxford University Press: Melbourne

—— 1973. *Fiji Indian Migrants*, Oxford University Press: Melbourne

Girvan, N. 1976. 'White Magic: The Caribbean and Modern Technology'. Papers presented at a Seminar on Caribbean issues related to UNCTAD IV, University of the West Indies: Mona (*Mimeographed*)

Goheen, M. 1993. 'Les champs appartiennent aux hommes, les récoltes aux femmes: accumulation dans la région de Nso', in P. Geschiere and P. Konings (eds), *Itinéraires d'accumulation au Cameroun*, Karthala: Paris, pp. 241–71

Goveia, E. 1965. *Slave Society in the British Leeward Islands at the End of the Eighteenth Century*, Yale University Press: New Haven and London

Graham, G.E. and Floering, I. 1984. *The Modern Plantation in the Third World*, St Martin: London

Greaves, I.C. 1959. 'Plantations in World Economy', in *Plantation Systems of the New World*. Papers and Discussion Summaries of the Seminar Held in San Juan, Puerto Rico, Pan American Union/Research Institute for the Study of Man: Washington DC

Green, L. 1925. *The Planter's Book of Caste and Custom*, Blackfriars House: London and The Times of Ceylon Co. Ltd.: Colombo

Guha, A. 1977. *Planter Raj to Swaraj: Freedom Struggle and Electoral Politics in Assam, 1826–1947*, Indian Council of Historical Research: New Delhi

Hall, D. 1953. 'The Apprenticeship Period in Jamaica, 1834–1838', *Caribbean Quarterly* 3:3

Harewood, J. 1975. *The Population of Trinidad and Tobago*, CICRED Series: Port of Spain

Hartman, H. 1981. 'The Unhappy Marriage of Marxism and Feminism: Towards a More Progressive Union', in L. Sargent (ed.), *Women and Revolution: The Unhappy Marriage of Marxism and Feminism*, Pluto Press Ltd.: London

Hatchard, J. 1837. *Negro Apprenticeship in the Colonies: A Review of the Report of the Select Committee of the House of Commons*, John Hatchard & Son: London

Heidemann, F. 1992. *Kanganies in Sri Lanka and Malaysia*, Miinchen: Anacon

Higman, B.W. 1976. *Slave Population and Economy in Jamaica, 1807–1834*, Cambridge University Press: Cambridge

—— 1980. The Jamaican Censuses of 1844 and 1861, Social History Project, Department of History, Mona, pp. 31–9 (*Mimeographed*)

Hoefte, R. 1987. 'Female Indentured Labour in Suriname: For Better or Worse?', *Boletin de Estudios Latinoamericanos y del Caribe* 42: 5–70

IOR 1882. India Office Records, V/ 24/ 1210. London

ILC Report 1859. *Report of the Immigrant Labour Commission for the half-year ending 30th June, 1859*, Immigrant Labour Commission: Colombo

ILO 1966. *Plantation Workers: Conditions of Work and Standard of Living*, ILO Publications: Geneva

Isaacman, A. 1990. 'Peasants and Rural Social Protest in Africa', *African Studies Review* 33 (2): 1–120

Jackson, S. 1978. 'Hausa Women on Strike', *Review of African Political Economy* 13: 21–36

Jain, D. 1980. *UPASI Study Report. A Preliminary Survey*, Institute of Social Studies Trust: New Delhi

Jain, R.K. 1970. *South Indians on the Planatation Frontier in Malaya*, Yale University Press: New Haven and London

—— 1986. 'Freedom Denied?: Indian Women and Indentureship', *Economic and Political Weekly* 21(7): 316

Jain, S. 1982. 'Women Workers on a Tea Garden in Assam: A Social Anthropological Study of the Nimari Labouring Community', *Ph.D. Thesis*, Jawahalal Nehru University

—— 1983. 'Tea Gardens in Assam: Patterns of Recruitment, Employment and Exploitation of Tribal Labourers', *Social Action* 33: 262–84

—— 1986. 'Sex Roles and the Dialectics of Survival and Equality', in L. Dube, E. Leacock and S. Ardener (eds), *Visibility and Power: Essays on Women in Society and Development*, Oxford University Press: New Delhi, pp. 158–93

—— 1988. *Sexual Equality : Workers in an Asian Plantation System*, Sterling Publishers Private Limited: New Delhi

—— 1991. 'Acculturation Process in an Assam Tea Garden', *The Eastern Anthropologist* 44(1): 13–43

JA IL 1720–71. Jamaican Archives, Inventories Libers. Kingston

JAR 1920. *Jamaica Annual Report*, pertaining to the report of F.N. Issacs, Protector of Immigrants, Jamaica Annual Department Reports: Kingston

JCR 1871–1943. *Population Census of Jamaica* from 1871 to 1943, Kingston

JT 1915. *Jamaica Times*, 8 May 1915 with an extract from a report by Messrs. Lal and McNeil

Jayaraman, R. 1975. *Caste Continuities in Ceylon: A Study of Social Structure of Three Plantations*, Popular Prakashan: Bombay

Jayawardena, C. 1963. *Conflict and Solidarity in a Guyanese Plantation*, The Athlone Press: London

—— 1975. 'Farm, Household and Family in Fiji Indian Rural Society', (Part One and Part Two) *Journal of Comparative Family Studies*

Jayawardena, L. 1963. 'The Supply of Sinhalese Labour to Ceylon Plantations (1830–1930): A Study of Imperial Policy in a Peasant Society', Ph.D. thesis, University of Cambridge

Jha, J.C. 1975. 'The Background to the Legalization of Non-Christian Marriages in Trinidad and Tobago', paper presented to the Conference of East Indians in the Caribbean, St. Augustine (*Mimeographed*)

—— (not dated) I The Beginnings of Indian Immigration to West Indies. (*Mimeographed*)

—— (not dated) II Indentured Indian Immigration 1835–1917. (*Mimeographed*)

Kaberry, P.M. 1952. *Women of the Grassfields: A Study of the Economic Position of Women in Bamenda, British Cameroons*, HMSO: London

Kane, F. 1977. 'Femmes prolétaires du Sénégal, à la ville et aux champs', *Cahiers d'Etudes Africaines* 17 (1): 77–91

Kannabiran, K. 1989. Review of *Sexual Equality: Workers in an Asian Plantation System* by Shobhita Jain. *Contributions to Indian Sociology* (n.s.) 23(2): 370

Kemp, C. 1985. 'The Planting Ideology in Sri Lanka: A Brief Enquiry into its Form, Content and Significance in Relation to the Tea Industry before Nationalisation', Paper presented to the South Asia research conference, School of Oriental and African Studies, University of London

Kondapi, C. 1951. *Indians Overseas in Colonial Territories,* Oxford University Press: New Delhi

Konings, P. 1993. *Labour Resistance in Cameroon: Managerial Strategies and Labour Resistance in the Agro-Industrial Plantations of the Cameroon Development Corporation,* James Currey: London

—— 1995a. 'Plantation Labour and Economic Crisis in Cameroon', *Development and Change* 26 (3): 525–49

—— 1995b. *Gender and Class in the Tea Estates of Cameroon,* Aldershot: Avebury (African Studies Centre Research Series no. 5)

Krishnamurty, J.K. (ed.) 1989. *Women in Colonial India: Essays on Survival, Work and the State*, Oxford University Press: New Delhi

Kurian, R. 1982. *Women Workers in the Sri Lanka Plantation Sector: An Historical and Contemporary Analysis,* International Labour Office: Geneva

Lal, B.V. 1983. 'Girmitiyas: The Origins of the Fiji Indians', *The Journal of Pacific History*

—— 1985a. 'Veil of Dishonour – Sexual Jealousy and Suicide on Fiji Plantations', *The Journal of Pacific History*

—— 1985b and 1989. 'Kunti's Cry: Indentured Women on Fiji Plantations', *The Indian Economic and Social History Review* (later published in J.K. Krishnamurty 1989. *Women in Colonial India: Essays on Survival, Work and the State*, Oxford University Press: New Delhi

Lal, C. and McNeil, J. 1915. *Report on Emigration from India,* HMSO Cmd 7744: London, pp. 223–320

Laurence, K.O. 1971. *Immigration into the West Indies in the Nineteenth Century*, Caribbean Universities Press: Barbados

—— 1994. *A Question of Labour: Indentured Immigration into Trinidad and British Guiana, 1875–1917,* Ian Randle Publishers: Kingston

LC Report 1908. *Report and Proceedings of the Labour Commission* (headed by Sir Hugh Clifford), Government of Ceylon: Colombo

Leacock, E. 1986. 'Women, Power and Authority', in L. Dube, E. Leacock and S. Ardener (eds), *Visibility and Power: Essays on Women in Society and Development,* Oxford University Press: New Delhi, pp. 107–35

Lewis, M.G. 1834. *Journal of a West India Proprietor, Kept During a Residence in the Island of Jamaica*, John Murray: London

Lewis, W. Arthur 1969. *Aspects of Tropical Trade 1883–1965*, Almquist and Wisell: Stockholm

Lieten, G.K. and Nieuwenhuys, O. 1989. 'Introduction: Survival and Emancipation', in G.K. Lieten, O. Nieuwenhuys and L. Schenk-Sandbergen (eds), *Women, Migrants and Tribals: Survival Strategies in Asia,* Manohar: New Delhi

Lieten, G.K., Nieuwenhuys, O. and Schenk-Sandbergen, L. (eds), 1989. *Women, Migrants and Tribals : Survival Strategies in Asia,* Manohar: New Delhi

Loewenson, R. 1992. *Modern Plantation Agriculture: Corporate Wealth and Labour Squalor,* Zed Books: London/New Jersey

Long, Edward 1774. *The History of Jamaica.* 3 vols, T. Lowndes: London

Look Lai, W. 1993. *Indentured Labor, Caribbean Sugar: Chinese and Indian Migrants to the British West Indies, 1838–1918,* The Johns Hopkins University Press: Baltimore

Lopez-Gonzaga, V. 1985. 'The Sugarcane Workers in Transition: The Nature and Context of Labor Circulation in Negros Occidental' (*Mimeographed*), La Salle – Social Research Centre: Bacolod, pp. 33–5

—— 1986. 'Crisis in Sugarlandia : The Planters' Differential Perceptions and Responses and their Impact on Sugarcane Workers' Household', Final Report Submitted to the Visayas Research Consortium and the Phil. Social Science Council (*Mimeographed*), La Salle – Social Research Centre, Bacolod

—— 1987. 'Capital Expansion, Frontier Development and the Rise of Monocrop Economy in Negros (1850–1895)' (Occasional Paper No. 1, *Mimeographed*) La Salle – Social Research Centre: Bacolod

Mackintosh, M. 1989. *Gender, Class and Rural Transition: Agribusiness and the Food Crisis in Senegal,* Zed Books: London/New Jersey

Mandelbaum, D.G. 1970. *Society in India: Change and Continuity,* University of California Press: Berkeley

Mangahas, M. 1985. 'Rural Poverty and Operation Land Transfer in the Philippines', in *Strategies for Alleviating Poverty in Rural Asia,* Bangladesh Institute for Development Studies: Daca

Marcelo, R. 1986. *The Manila Chronicle,* 22 September

Mathurin, L. 1974. 'A Historical Study of Women in Jamaica, 1655–1844', Ph.D. Thesis, University of the West Indies, Mona, Jamaica

Mayer, A.C. 1973. *Peasants in the Pacific: A Study of Fiji Indian Rural Community,* Routledge and Kegan Paul: London

Mbilinyi, M. 1988. 'Agribusiness and Women Peasants in Tanzania', *Development and Change* 19 (4): 549–83

McCoy, Alfred 1983. The Philippine Sugar Industry Under Martial Law: World Market Pressure and Farm Mechanization. (*Mimeographed*)

McNeil, J. and Lal, C. 1915. *Report to the Government of India on the Condition of Indian Immigrants in Four British Colonies and Surinam,* T. Fisher, Unwin: London

Mies, M. 1980a. *Towards a Methodology of Women's Studies,* Occasional Paper. Institute of Social Studies: The Hague

—— 1980b. *Indian Women and Patriarchy,* Concept Publishing Co.: New Delhi

Miles, R. 1987. *Capitalism and Unfree Labour: Anomaly or Necessity?,* Tavistock: London

Mintz, S.W. 1956. 'Cañamelar: The Sub-culture of a Rural Sugar Plantation Proletariat', in J.H. Steward, et al. *The People of Puerto Rico: A Study in Social Anthropology,* University of Indiana Press: Urbana, pp. 314–417

—— 1957. 'The Plantation as a Socio-cultural Type', in *Plantation Systems of the New World,* Washington D.C.: Pan American Union (Social Science Monographs VII)

—— 1974. 'The Rural Proletariat and the Problem of Rural Proletarian Consciousness', *Journal of Peasant Studies* 13: 291–325

—— 1985. *Sweetness and Power, The Place of Sugar in Modern History*, Donnely and Sons Co.: Virginia

Mintz, S.W. and Wolf, E.R. 1950. 'An Analysis of Ritual Co-parenthood (compadrazgo)', *Southwestern Journal of Anthropology* 4

MMSA 1914–23. *Methodist Missionary Society of Australasia Misc. papers and M/33 Folder.* Fiji National Archives: Suva

Morton, S.E. 1916. *John Morton of Trinidad,* Westminister Co.: Toronto

Naidu, V. 1980. *The Violence of Indenture in Fiji,* Suva World University Service and University of the South Pacific: Suva

Narsey, W.L. 1979. 'Monopoly Capital, White Racism and Superprofits: A Case Study of CSR', *Journal of Pacific Studies*

NEDA 1982. *Philippine Statistical Yearbook,* National Economic And Development Authority: Manilla

Nkwi, P.N. 1985. 'Traditional Female Militancy in a Modern Context', in J.C. Barbier (ed.), *Femmes du Cameroun: Mères pacifiques, femmes rebelles,* Orstom/Karthala: Paris, pp. 181–91

O'Barr, J. 1984. 'African Women in Politics', in M.J. Hay and S.B. Stichter (eds), *African Women South of the Sahara,* Longman: London/New Jersey, pp. 140–55

Obbo, C. 1980. *African Women: Their Struggle for Economic Independence,* Zed Press: London

Padilla, E. 1957. 'Contemporary Socio-rural Types in the Caribbean Region', in V. Rubin (ed.), *Caribbean Studies: A Symposium,* Institute of Social and Economic Research: Jamaica

Parliamentary Papers 1789, 1832. *British Parliamentary Papers,* relating to West Indian Slavery. India Office Records: London

Parpart, J.L. 1988. 'Women, Work and Collective Labour Action', in R. Southall (ed.), *Labour and Unions in Asia and Africa,* Macmillan: Basingstoke, pp. 238–55.

Patterson, O. 1967. *The Sociology of Slavery,* McGibbon and Kee: London

Pittin, R. 1984. 'Gender and Class in a Nigerian Industrial Setting', *Review of African Political Economy* 31: 71–81

Popovic, A. 1965. 'Ali Ben Mohammed et la révolte des esclaves à Basra', doctoral dissertation, Universite de Paris

POSG 1903. *Port of Spain Gazette – Column, 'The Police Courts'* 25.4.1903

Post, K. 1978. *Arise Ye Starvelings: The Jamaican Labour Rebellion of 1938 and its Aftermath,* Martinus Nijhoff: The Hague

Presley, C.A. 1986. 'Labor Unrest among Kikuyu Women in Colonial Africa', in C. Robinson and I. Berger (eds), *Women and Class in Africa,* Africana Publishing Company: New York/London, pp. 255–73

Ramachandran, S. 1994. *Indian Plantation Labour in Malaysia,* S. Abdul Majeed & Co. for Institute of Social Analysis, Malaysia

Ramaswamy, V. 1993. 'Women and Farm Work in Tamil Folk Songs', *Social Scientist* 21 (9,10,11): 113–29

Ramnarine, T. 1980. 'Indian Women and the Struggle to Create Stable Marital Relations on the Sugar Estates of Guiana during the period of Indenture, 1837–1917'. Paper presented to the 12th Conference of Caribbean Historians, St. Augustine (*Mimeographed*)

Reddock, R. 1984. 'Women, Labour and Struggle in Twentieth Century Trinidad and Tobago', Doctoral Dissertation, University of Amsterdam

—— 1985. 'Freedom Denied: Indian Women and Indentureship in Trinidad and Tobago, 1845–1917', *Economic and Political Weekly* 20 (43): WS - 79 - 87

—— 1993. 'Douglarization and the Politics of Gender Relations', in R. Deosaran, N. Mustapha and R. Reddock (eds), *The Contemorary Caribbean,* University of the West Indies: St. Augustine

—— 1994. *Women, Labour and Politics in Trinidad and Tobago: A History,* Zed Books: London and New Jersey.

Roberts, G.W. 1957. *The Population of Jamaica*, Cambridge University Press: Cambridge

Roughly, T. 1823. *The Jamaica's Planter's Guide*, London

Rubin, V. (ed.) 1957. *Caribbean Studies: A Symposium*, Institute of Social and Economic Research: Jamaica

Safa, H.I. 1979. 'Class Consciousness among Working Class Women in Latin America: A Case Study in Puerto Rico', in R. Cohen, P.C.W. Gutkind and P. Brazier (eds), *Peasants and Proletarians: The Struggles of Third World Workers,* Hutchinson and Co: London, pp. 441–59

Sajhau, J.P. and Von Muralt 1987. *Plantation Workers,* International Labour Office: Geneva

Samarveera, V. 1981. 'Masters and Servants in Sri Lanka Plantations: Labour Laws and Labour Control in an Emergent Export Economy', *The Indian Economic and Social History Review* 18-2: 123–58

Sanderson, Commission 1910. *Report on Emigration from India to the Crown Colonies and Protectorates Parts I and II and Minutes of Evidence and Papers laid before the Committee.* HMSO, Cmd 5192, 5193 and 5194: London

Selby, et al. 1987. 'Battling Urban Poverty from Below: A Profile of the Poor in Two Mexican Cities', *American Journal of Sociology* 87: 419–20

Selyman, E. and Johnson, A. 1948. *Encyclopedia of the Social Sciences.* Vol. 11, Macmillan: New York

Shanklin, E. 1990. '*Anlu* Remembered: The Kom Women's Rebellion of 1958–61', *Dialectical Anthropology* 15(3): 159–81

Shepherd, V. 1986. 'From Rural Plantations to Urban Slums', *Immigrants and Minorities* 5: 130–43

—— 1993. 'Emancipation through Servitude?: Aspects of the Condition of Indian Women in Jamaica, 1845–1945', in H. Beckles and V. Shepherd (eds), *Caribbean Freedom: Society and Economy from Emancipation to the Present,* Ian Randle: Kingston, pp. 245–50

—— 1994. *Transients to Settlers: The Experience of Indians in Jamaica, 1845–1950*, Peepal Tree Press: Leeds & Centre for Research in Asian Migration, University of Warwick.

—— 1995. 'Gender, Migration, Indentureship and Settlement', in V. Shepherd, B. Bailey, B. Brereton (eds), *Engendering History: Caribbean Women in Historical Perspective,* Ian Randle Publications: Kingston

Shepherd, V., Bailey, B. and Brereton, B. (eds) 1995. *Engendering History: Caribbean Women in Historical Perspective*, Ian Randle Publications: Kingston

Shiels, R. 1969. 'Indentured Immigration into Trinidad 1891–1916'. B. Lit. thesis, University of Oxford

Siddique, M.A.B. 1990. *Evolution of Land Grants and Labour Policy of Government: The Growth of the Tea Industry in Assam 1834–1940*, South Asian Publishers: New Delhi

Silva, K.T. 1979. 'The Demise of Kandyan Feudalism', in B.M. Morrison, M.P. Moore and M.U. Ishar Lebbe (eds), *The Disintegrating Village: Social Change in Sri Lanka*, Lake House Investments: Colombo, pp. 43–70

SLNA 1840–1874. *Sri Lanka National Archives.* Government of Ceylon: Colombo (The reference in the text is given by the relevant file)

Stichter, S.B. 1985. *Migrant Laborers*, Cambridge University Press: Cambridge

Stichter, S.B. and Parpart, J.L. (eds) 1988. *Patriarchy and Class: African Women in the Home and the Workforce*, Westview Press: Boulder, Colorado

Sturge, J. and Harvey, T. 1838. *The West Indians in 1837*, Adams & Co.: Hamilton, London

Sutherland, W.M. 1984. 'The State and Capitalist Development in Fiji', Ph.D. Thesis, University of Canterbury

Sutton, C. and Makiesky-Barrow, S. 1977. 'Social Inequality and Sexual Status in Barbados' in A. Schlegel (ed.), *Sexual Stratification: A Cross-Cultural View*, Columbia University Press: New York, pp. 129–56

SYJ 1981. *Statistical Yearbook of Jamaica.* Kingston

Thompson, E.T. 1959. 'The Plantation as a Social System', in *Plantation Systems of the New World*, Pan American Union: Washington D.C.

Tiffin, M. and Mortimer, M. 1990. *Theory and Practice in Plantation Agriculture*, Westview: London

Tinker, H. 1974. *A New System of Slavery: The Export of Indian Labour Overseas 1830–1920*, Oxford University Press: London

Tyson, J.D. 1939. *Report on the Condition of Indians in Jamaica*, Government Press: Simla

Uberoi, P. 1987. Review of *Visibility and Power: Essays on Women in Society and Development.* Edited by L. Dube, E. Leacock and S. Ardener, *Contributions to Indian Sociology* (n.s.) 21(2): 388–90

Underhill, E. 1970. *The West Indies: Their Social and Religious Condition*, Negro Universities Press: Connecticut (Originally published by Jackson, Waltford & Hodder, London in 1862)

Vaughan, M. and Chipande, G.H.R. 1986. Women in the Estate Sector of Malawi: The Tea and Tobacco Industries. (*Mimeographed*) Geneva: ILO, WEP 10/WP. 42.

Webster's Third International Dictionary 1971.

Weller, J.A. 1968. *The East Indian Indenture in Trinidad*, Institute of Caribbean Studies, Caribbean Monograph Series, No. 4: Rio Piedras

Wilkinson, A. 1989. *Big Sugar*, Knopf: New York

Williams, E. 1952. *Documents on British West Indian History, 1807–1833*, Historical Society of Trinidad and Tobago, Port of Spain: Trinidad

—— 1964. *A History of the People of Trinidad and Tobago*, Andre Deutsch: London

WIRC 1898. *West Indian Royal Commission Report.* London

Wolf, E.R. and Mintz, S.W. 1957. 'Haciendas and Plantations in Middle America and the Antilles', *Social and Economic Studies* 6(3): 380–412

Wood, D. 1968. *Trinidad in Transition: The Years After Slavery*, Oxford University Press: Oxford

Index

absenteeism, 30, 39, 157, 161
alcohol, 79, 143
  alcoholic drinks, 81, 125
  alcoholism, 143
  confirmed a+lcoholics, 125
  expenditure, 143
    toddy shop, 9
    liquor, 83
authoritarian structure, 86, 109
  managerial and supervisory, 159
authority, 10, 52, 70, 110, 112, 114, 121, 161
  extra-legal, 110
  male, 5, 15, 109, 111, 115, 118–20
    kangani/sardar, 12, 71, 115
      work-related, 115
    patriarchal, 118, 126
  traditional, 80
    caste, 40
    pundits, 40
    parental, 117
autonomy, 29, 38, 45, 48
  curtailment, 38
  degree, 43
  women's struggle, 11, 38
  Indian women, 29

benefits, 8, 84, 113, 127, 130, 156
  fringe, 157, 159, 162, 164
  maternity, 84, 85, 127, 148
    allowance, 25
    leave, 15
bride-price, 41, 117, 120

cash, 101, 115, 120, 132, 144
  access, 115
  crops, 9, 18, 26
  extra, 116
  generating, 139
  income, 39

loan, 136
plucking, 73, 74, 86n1
wages, 9, 117, 156
caste(s), 11, 13, 52, 62, 64, 67, 69, 75, 111, 116, 118, 125
  authority, 81
  casteism, 86
  differences, 78–9
  endogamy, 38–40
  hierarchy, 13, 76, 80
    sanskritization, 76–9
  high, 40, 79, 81
  identity(ies), 77, 118
  ideology(ies), 12, 77
  impurity and pollution, 39
  lower, 41, 42, 44, 76, 78, 81, 117
  mobility, 42
  segregation, 79
  status, 11, 42
    higher, 68, 78, 81
  sub-castes, 13, 77, 78
  system, 40
    quasi-caste, 77
  twice-born, 39
  values, 76, 80
child bearing, 22, 108
childcare, 5, 60, 99, 138, 141, 157
child-minding, 122, 127
  baby-sitting, 141
children, 22–4, 26, 33–4, 51, 56, 59, 72–5, 80, 84, 93–4, 98, 99, 100, 108, 117–18, 120, 127n10, 153, 157
  birth, 95, 122
  contribution, 15
  death rate, 57
  female, 41, 79, 122
  freed, 26
  grandchildren, 131
  Indian immigrants', 91

Indian-Jamaican, 102
loss, 56
male, 79
slave, 26
subsistence, 11
unmarried, 121
working, 116
class(es), 109, 110
analysis, 89
formation, 9
gender, 52
inequalities, 12, 15
'labour', 9, 117
landless labouring, 115
landowners, 30
middle, 113
planter, 101, 102
prostitute, 52
relations, 53
social, 129
stratification, 8
structure, 9
struggle, 53, 164
supervisory and managerial, 110
upper, 110
white indentured servants, 19
working, 34, 62, 130
white women, 90
colonial state, 11, 37–8, 42, 48–9, 51, 69, 86
administrative machinery, 109
colonialism, 51
government, 9, 33, 49, 69
official(s), 49, 58–9
planter, 111; *see also* plantation(s), colonial
powers, 158
coolie(s), 5. 49, 68, 123
identity, 111
Indians, 5
transportation, 54
creche(s), 9, 73–4, 84, 91, 98–9, 127
attendants, 73, 84, 127n10
creole society, 38
attitudes towards women, 24
historian, 22
socio-economic ladder, 19

decision-making, 43, 81, 108, 117, 158
active role, 120

independence, 117
power, 121
division of labour, 10, 50, 63–4, 75, 80, 82, 90,
108, 112, 115, 122, 123
gender, 90, 97
household, 15, 75, 121, 138
sexual, 1, 18, 33, 55, 58–9, 72, 96
divorce, 120
ban, 51
divorcee, 127, 153
domestic arrangement, 37
chores, 138
duties, 63–4
economy, 11, 26, 37, 48
food, 27
group, 120
helpers, 131, 145
ideology, 48
labourers, 54
market, 156
needs, 54, 60
servants, 13, 96, 101, 103
service, 97, 101, 140
work, 37
domestic labour debates, 53, 62
domestication process, 31
domestics, 105
planters' houses, 19
occupational rating, 20
dowry, 41, 117

education, 51, 83–4, 154
children's, 146
compulsory, 84
educational level, 102, 154
standard, 103
English, 69
low level, 15, 132, 147
egalitarian, 108
attitude, 77
egalitarianism, 108
values, 108
emancipation of women, 11, 52–3, 64, 90
equality, 109
access, 126
ideology, 125
inequality, 14, 125
sexual, 14

social, 14
European(s), 50, 69
    colonization, 2
    colonizers, 96
    manufactured goods, 6
    managerial roles, 9
    patriarchy, 90
    planters, 54, 109
    powers, 6
    privileged position, 69

family(ies), 8, 29, 42, 53, 58, 59, 62, 69, 72–3,
        90, 96, 100, 108, 111, 114, 117, 120
    authority, 10, 115
        male dominated, 15
        parental, 117
    control of purse, 116
    debts, 85
    economic unit, 108
    elders, 153
    emigration, 93, 94
    European conjugal, 48
    extended, 121, 131, 138
    farms, 153
    formation, 110, 116, 118
    group(s), 69
    head, 15, 121
    Indian, 44, 45, 54, 91
        patriarchal, 38, 42, 48
    inheritance, 129
        property, 42
    members, 11, 50, 75, 79, 116; *see also*
        migrant, family(ies)
    nuclear, 38, 63, 120, 122, 138
    of orientation, 119, 125
    organization, 42
    ownership, 8
    planning, 14, 85
    plots, 7, 63, 85
    reproduction, 153, 163
    size, 15
    stability, 91
    Sudra, 79
    system, 38, 124
    ties, 100
    wages, 14
    working members, 83
famines, 67

Indian, 30
fertility, 96
    rates, 91, 99
        lowest, 24
festivals, 79, 87n4
food, 67, 73–5, 78, 84–5, 87n5, 91, 133, 145
    crops, 26, 37, 45
    cost, 57
    economy, 27
    expenditure, 143
    nutritious, 135
    producers and distributors, 26
    providers, 138

gang(s), 20, 50, 58, 61, 71, 96
    field, 22, 25
    labour, 78, 161
    placement, 97
    tipping, 71
    weeding, 97
gender, 14, 52, 90–6
    allocation of work, 15
    composition, 161; *see also* division of labour
    assumptions, 21
    differences, 96
    dimension, 21
    discrimination, 10, 96, 103
    equity, 107
    ideology, 13, 90
    immigration policy, 103
    perspective, 89
    relations, 12, 14, 89, 107, 110, 121
        parity, 123
    roles, 1, 12, 90, 111, 121
    stereotypes, 124

health, 23, 75
    facilities 5
    hospitals, 51, 84
hierarchy, 68, 70, 76–7; *see also* caste; division
        of labour; power, relations
    occupational and social, 110
    residential pattern, 69
    respect for, 12
    slave, 19
    system, 12, 69
household organization, 108, 121, 123, 131
    activities, 108, 121

adaptability, 135
breadwinner, 132
budget(ing), 125, 134
chores/duties/tasks, 34, 53, 74, 75, 114, 116,
    120, 123
common type of, 121
composition, 116, 122
    size, stage, 116, 132
consumption, 137
duties, 34
expenditure, 115
flexibility, 121, 131, 133, 147
head, 83, 153, 163
    female-headed, 118
income, 115, 121, 133, 135, 138, 148
    single-income earner, 121, 136
labour, 60
management, 129
needs, 69, 153
power within, 35
production and reproduction, 71–5
property, 144
role, 71, 72
sharing, 145
single-parent, 131
studies of, 147
survival, 129, 131, 137
    strategies, 111
tribal, 111
    structure, 80
housing, 5, 78, 84, 86n3, 120, 127n10, 157
barracks, 38, 39, 50, 69, 86, 99
'line(s)', 50, 73, 78, 84, 86n3, 101, 114,
    116–18, 120, 121, 123, 125, 126n4
requirements, 45

ideology(ies), 10, 64, 75, 103
alternative, 107
breadwinner, 103
contemporary, 1
male dominance, 155
plantation, 12, 69, 70
racist, 10, 21
Victorian, 94
    white supremacy, 86; *see also* caste,
        equality, gender, patriarchy
immorality, 36, 44, 93
plantation, 50, 51

women, 61, 62
income, 36, 68, 75, 115–17, 153, 156
adequate, 163
additional, 83, 157
cash, 22, 39
pooling, 125
real, 164
rights over, 83
    see also cash, access
source, 155
variation, 82
wage, 153, 164
indentured labour, 5, 13, 30, 49, 51, 97
immigrant, 99
Indian, 29
servitude, 4, 13
women, 51, 52, 54–5, 58–61, 90–1, 97
indentureship, 11, 14, 29, 30–1, 37, 40, 45, 89,
    90–2, 95–6, 103
continuation, 99
five-year period, 54, 56
history, 62
length and conditions, 36
system, 3–4, 33–4, 50–1, 57, 60, 99

kangani, 12, 71, 73, 77, 80, 85, 86, 115, 127n7
de facto leader, 69, 76
head, 78
status, 12
sub-*kangani*, 78
system, 12, 68
    institutionalization, 115
    study, 12
kinship, 62, 78, 114, 121
kin-groups, 68, 76
ties, 76, 78

labour, 5, 49, 137; *see also* indentured labour;
    management, labour
categories, 34
agricultural, 63, 91, 96, 101, 158
casual, 9
field, 19, 23, 45, 50, 89, 90, 97
local, 2
manual, 10, 21
rural, 17
permanent, 92
skilled, 6

control, 6, 45, 69–70, 72, 80, 163
costs, 57
displacement, 130
hired, 76
household, 72
market, 14, 103
migrant, 96
muster, 10
offences, 60
organization, 1, 4
policy, 14
power, 24, 26, 152
process, 15, 71, 151–2, 158–9
protests, 163
recruitment, 14, 89
resistance, 159, 164
shortages, 25, 99
reproduction, 10
    see also slave, labour
statistics, 63
supply, 5, 63, 67
unpaid, 153; see also wage(s), labour
welfare, 10
land, 30, 49, 67, 100, 131, 135, 148
commutation grants, 101
food production, 156
garden plots, 100, 101, 126n5
government, 9, 14
grants, 101
hacienda, 8
landholder(s), 91, 93, 96, 103
    attitude, 90
    landlord and the worker, 146
marginal, 101
ownership, 67
    landowners, 49, 92
paddy, 9
use, 5, 6

management, 69, 78, 83, 85–6, 112–13, 151,
    155, 157, 159, 164
and union leadership, 159, 160, 162
circles, 158
controls, 159
estate (plantation), 12, 38, 69, 70, 75, 80, 84,
    85, 118, 121, 152, 156
financial, 15
interest, 80

labour, 77
    henchmen, 114
male-dominated, 155
manager(s), 50, 51, 55, 60, 99
    estate (garden), 9, 12, 22, 114, 118, 161
general, 110
white, 9
marriage(s), 38, 40, 43, 52, 58, 94, 154
age at, 119
break up, 120
ceremony, 79
conditions, 50
dissolution, 153
forms, 10
    arranged by parents, 117
    child, 41, 119
    common-law (rajikhushi), 117, 121
    endogamy, 40
        outside endogamous group, 118
first, 117, 120
Hindu, 40
hypergamous, 78
inter-caste, inter-tribal, inter-ethnic, 119
legal, 117
incompability, 118
indissolubility, 120
institution, 62, 64
marital status, 94
ordinance, 51
partners, 117
proposal, 117
residence after, 119
sanctity, 61
widow-remarriage, 39
    ban, 30
Vellalan, 87
medical care, 127n10, 156
examination, 56
facilities, 84, 157
malnutrition, 57
nutrition, 56, 58
    pregnancy, 56, 84
sickness, 56, 57
    measles epidemic, 49
test of fitness, 122
migrants, 11, 13, 33, 137, 143–4
illiterate, 9; see also indentured labour
labourers/workers, 30, 133

second and third generations, 111
Indian, 30
missionary(ies), 50, 51, 103
  church, 155
  Methodist Mission, 51
  mission schools, 51
  Presbyterian, 42
mobility, 110; *see also* caste, mobility
  occupational, 113
    lower to upper classes, 110
    women workers, 124
    upward, 76
    lack of, 82
  social, 104
mortality, 57
  high, 10
  rate, 39, 93
mother(s), 37, 59, 60, 70, 84, 98, 138, 140
  'adopted', 23
  'carelessness and indifference', 57
  motherhood, 24, 27
  Negro, 26
  nursing, 74
  'picaninny', 24
  unmarried, 85
  working, 141

occupation(s), 31, 64, 75, 138, 154
  adult Indian, 31
  category(ies), 10, 63
    lowest paid, 33
    non-field, 97
    plucking, 124; *see also* tea, pluckers
    Sardar and Chowkidar, 112
    skilled, 19
    slave, 90
occupational norms, 18
  options, 25; *see also* mobility, occupational
  pattern, 101–2
  rating, 20
  remunerative, 100

patriarchal controls, 15, 152–5, 163
  opposition, 153
  thinking, 97
  (trade) union, 159
  values, 33
patriarchy, 15, 80, 120, 124

family, 121; *see also* family, patriarchal; *see also* ideology, patriarchy
  patterns, 84
  traditional, 86
plantation(s), 30, 41
  brutality, 60, 62
  colonial, 6, 109–10
  crops, 3, 7, 70
    banana, 3, 97, 98
    cocoa, coconut, 3, 11, 70
    cotton, 3, 5
    sugar, 2–3, 5, 18
    tea, 3, 7, 67, 70
  conditions, 58
  economy, 11, 29, 67
  enclaves, 2, 6, 109
  hierarchical, 50, 55
  industrial, 4, 5, 6, 163
  labour, 5, 153, 155, 165n1; *see also* production, factors, means, mode
  new style, 4
  New World, 17
  old style, 4
    slave, 4, 21
    traditional, 4–5
    'total institution', 1, 111
  origins, 2 -4
    tropical, 2
  'plantation community', 68
    Tamil, 80
  political economy, 21
  production, 31
  system, 4–7, 38–9, 53, 63, 76, 104
    agricultural estate, 7–8
    class structured, 4, 9, 121
    economic institution, 2, 4
  Southern USA, 5
  West Indies, 2, 3, 5
    Jamaica, 5, 89
planter(s),4, 9, 11, 34–5, 67–9, 90–1, 110; *see also* colonial, planter; European, planters
  absentee, 19
  Assam tea, 109
  brutality, 24
  class, 101–2
  dominance, 109
    dominated local legislature, 25

German, 151
hostility, 14, 24
households, 96
interest, 39
perception, 122
power, 86
sugar-cane, 130
power, 8, 12, 49, 76, 80, 163; *see also* decision-
making, power; labour, power; planter,
power; patriarchy
estate trade unions, 86
male, 59
abuse, 158; *see also kangani*, head
mystical, 162
political, 80
relations, 71
social, 68
structure(s), 81, 110
male, 85
struggles, 156
production, 70; *see also* plantation, production
capitalist, 52, 58
commodity, 89
cost, 92
distribution, 8
domestic and subsistence, 54
factors, 6, 8
food, 156, 157
women's, 158
market, 48
means, 158
methods, 63
mode, 2
capitalist, 54
peasant, 6, 31, 37, 45
subsistence, 37, 45, 54
process, 4, 5, 63, 111
relations, 4, 45
social, 109
surplus, 8
system(s), 62, 63, 75, 82
techniques, 71
proletarianism, 115
proletarianization, 89
proletariat, 27, 129
property, 40, 41, 79; *see also* family, property
absence, 99
owners, 19

slave, 22
prostitute(s), 50, 52, 58, 93, 155
prostitution, 14, 58, 93

race(s), 44, 69, 89, 97, 103
advantage of, 19
mixed-race, 44
racist, 63
ration(s), 5, 11, 91, 97, 135
cost, 39
free, 100
rice, 84
pregnant women, 35
subsidized, 116–17
system, 39
recruiters, 91, 93
recruitment, 30, 76, 89, 90, 152
labour, 14, 89
women, 30
method, 77
reproduction, 5, 24, 52, 58, 65n1, 70; *see also*
family(ies), reproduction
agents, 108
capacity, 10
capitalism, 53
cost, 11, 42
labour power, 35, 38, 45, 48
process, 5, 71
reproductive tasks, 54
social, 5
role, 50, 129, 147, 152, 163–4; *see also* gender,
roles
basic or ascribed, 111
child bearer, 152
critical, 135
crossing, 108, 123, 125
dominant, 131
female labour, 1
lack of rigidity, 108
managerial, 9
religion, 78
strategic, 129
subservient, 87n6
supervisory and clerical, 9
the Brahmin, 77

savings, compulsory scheme, 157, 162, 164
mechanisms, 116

sex-ratio(s), 40, 68, 93
  disproportionate, 11, 40, 42, 51
  male excess, 20
  unequal, 38
slave(s), 17; *see also* children, slave; hierarchy,
    slave; plantations, slave; property, slave
  deaths, 23
  distribution, 18
  emancipation, 3, 5, 18, 24
  pre-, 25
  post-, 27, 30
  entrepreneurship, 26
  ex-slaves, 25, 30, 31, 33
  field, 22
  female, 19, 22–6
    house, 20, 23
  labour, 5
  monetary value, 23
  multi-purpose work equipment, 22
  revolts and resistance, 5
  trade, 3, 20, 23
    British, 5
  women, 10, 37
slavery, 25, 30, 45, 50, 89, 97; *see also* wages,
    labour
  African, 89
  anti-slavery society, 42
    resistance, 99
  economics and politics of, 22
  end, 10
  Jamaican, 18
    post-slavery Jamaica, 90
  new system, 89
  West Indian, 20
status, 15, 33, 68, 79, 99, 111; *see also*
    caste, status; *kangani*, status; marriage,
    marital
  aspirations, 8
  decline, 23
  economic, 13, 75
  employment, 138
  financial, 8
  gender-related, 125
  improvement, 48
  index, 23
  inequalities, 15
  needs, 4, 8
  nutritional, 134

  positions, 110
  social, 124
    as workers, 76
    women, 22, 52
strikes, 35, 82, 159, 160–1
  wave of, 110
  strikers, 161
sugar, 2–3
  export, 49
  consumption, 130
industry(ies), 15, 64, 97, 101, 130, 147
  market, 45
  plantation(s)/estate(s), 2, 10–11, 14, 20, 31,
      37, 44, 49–50, 52, 57, 97, 102
  production, 49, 53, 130, 148
  workers, 57
suicide(s), 3, 36, 45, 55, 62, 120
  male, 62
  rates, 45
survival, 15, 64, 129, 137, 142, 147, 153, 159;
    see also family, survival, household,
    survival
  adaptive approach, 109
  day-to-day, 110
  mechanisms, 109
  strategies, 1, 108, 109, 113
    adaptive, 137
    adjustment, 144

tea, 7, 14, 151
  cultivation, 113
  China and Japan, 3
  Anglophone Cameroon, 15, 151
  Assam, 6, 9, 12, 14, 15, 107, 110
  industry, 69, 108, 113, 118
  pluckers, 72, 115, 123, 124, 151–2, 156,
      158
  production, 113
  Sri Lanka, 7, 12, 115
  theft of, 163
trade union(s), 81–3; *see also* power, estate
    trade unions
  leadership, 159, 160
  militant, 159
  movement, 81
  patriarchal structure, 159
  state-controlled, 152, 157, 160
  women's involvement, 158

value(s), 59, 78, 81, 110; *see also* caste, values;
    egalitarian, values; patriarchal, values
  devalue, 34
  judgements, 43
  market, 23
  monetary, 23
  plantation labour, 155
  scarcity, 51, 58
  surplus, 153
  system, 81–2, 84
  traditional, 110
  Victorian, 11
  women's chastity, 58
    virginhood, 44

wage(s), 7, 9, 34, 55, 70, 80, 82, 91, 93, 98,
    148, 155, 157, 162; *see also* cash, wages;
    family, wages
  annual, 98
  basic, 156–7
  daily, 97
  deduction, 84
  depressed, 37
  differentials, 35, 50, 57, 97, 103
  discrimination, 97
  earning population, 102
  equal work, 96
  labour, 11–12, 30, 33, 37, 45, 48, 54, 90, 103
    indentured, 34
  male member to collect, 83
  market, 25
  rates, 57, 98
    average, 35, 37, 57
    low, 37, 58, 93, 130
    minimum, 34, 162
    weekly, 115
  strikes and protests, 35
  structure, 10–11, 13, 82
  survival, 115
  Trinidad, 30
  Wage Board, 82
  wage policy, 10
women's movement, 29; *see also* autonomy;
    reproduction, capitalism; power, male
  Caribbean women's experience, 29
  class-oriented struggles, 163
  history, 29
  myths about docility, 29

resistance, 52–3, 60–1
  male overseers, 11
  rejection, 25
  subversive and aggressive strategy, 27
women's oppression, 14, 53, 107, 125; *see also*
    creole, attitudes towards; planter, hostility
  antagonism, 24
  assault and battery, 62
    flogging, 24
    gaol, 60, 61
  limited choices, 19
  loose character, 51
  moral downfall, 51
  murders, 44–5
  reputation, 58
  subservience, 79
  violence, 86
    and brutality, 60, 85
women's position, 52, 54, 62, 71–2, 78, 86,
    107, 152, 157; *see also* role, women's;
    status, women; value(s)
  changed status of women 54
  chastity, 58
  contradictory, 152
  control, 76–7
    absence of, 115
  free, 33–4
  health, 23
  'husbandless', 153, 163
  illiteracy, 83
  inferior to men, 79
  integration, 9–15
  unpaid, 59
  marginality, 52
  overseer, 161
  respectable character, 36
  scramble for, 37
  services, 59, 60
  single, 59, 94
  technical capabilities, 21
  traditional, 81
  universal subordination, 107
  weak position, 103
  wives and mothers, 37; *see also* childbearing
    pregnancy and confinement, 42
women workers, 6, 72, 82, 84–5; *see also*
    mobility, women workers; tea, pluckers;
    trade union, women's involvement; wages

African, 151, 158
  African-Jamaican, 94, 101
  enslaved African, 89
  proportion in African countries, 7
Indian, 31, 97

indentured, 54
Indian-African, 31
'Indian Women Problem', 30
Philippine society, 129
sugar-cane, 142, 148